Methods in Molecular Biology™

Series Editor
John M. Walker
School of Life Sciences
University of Hertfordshire
Hatfield, Hertfordshire, AL10 9AB, UK

For further volumes:
http://www.springer.com/series/7651

Fungal Diagnostics

Methods and Protocols

Edited by

Louise O'Connor and Barry Glynn

Molecular Diagnostics Research Group, NCBES, National University of Ireland, Galway, Ireland

Humana Press

Editors
Louise O'Connor
Molecular Diagnostics Research Group NCBES
National University of Ireland
Galway, Ireland

Barry Glynn
Molecular Diagnostics Research Group NCBES
National University of Ireland
Galway, Ireland

ISSN 1064-3745 ISSN 1940-6029 (electronic)
ISBN 978-1-4939-5980-8 ISBN 978-1-62703-257-5 (eBook)
DOI 10.1007/978-1-62703-257-5
Springer New York Heidelberg Dordrecht London

Softcover re-print of the Hardcover 1st edition 2013

Printed on acid-free paper

Humana Press is a brand of Springer
Springer is part of Springer Science+Business Media (www.springer.com)

Preface

Invasive fungal infections are a significant cause of morbidity and mortality. Over the past decade there has been a concerted effort to develop reliable methods for the detection of such infections. Early diagnosis of IFI is critical, allowing timely administration of appropriate antifungal therapy. The increased use of antifungal drugs has led to a parallel emergence of resistant and less-susceptible strains. This emergence has increased the importance of prompt and accurate identification of the causative agents of fungal infections. Rapid diagnosis is perhaps the most critical factor in ensuring a positive patient outcome. In this respect it is our hope that this volume may provide ideas for those intending to introduce novel technologies for fungal detection into their laboratories.

Key factors in introducing any new technology into a diagnostic setting include ease of use and rapid turnaround time without compromising sensitivity and specificity. To this end, we have focused on including in this volume methods which offer these characteristics with the potential to be adapted in the routine diagnostic setting.

While focusing specifically on fungal detection in clinical settings, the methods described are applicable to all areas utilizing fungal diagnostics including, environmental testing, agriculture and food production, and veterinary diagnostics. In addition, some of the methods described represent a significant practical demonstration of state-of-the-art molecular methods which are suitable for detection of non-fungal microorganisms and infectious agents. Included in this volume also are comprehensive reviews of fungal infections, a review of commercially available systems, as well as chapters concerned with sample preparation and a broad range of interesting methods for detection.

We would like to thank Professor John Walker and David Casey of Humana Press for this opportunity to become involved in this worthwhile project. Sincere thanks to the authors who gave of their own time to enlighten the broader scientific community of their expertise in this area of biology.

Galway, Ireland

Louise O'Connor
Barry Glynn

Contents

Contributors

JOANA BARBOSA • *Department of Microbiology, Faculty of Medicine, Cardiovascular Research Unit, University of Porto, Porto, Portugal; Escola Superior de Saúde Jean Piaget/ Vila Nova de Gaia, Gulpilhares, Vila Nova de Gaia*

JONAS BLOMBERG • *Department of Medical Sciences, Clinical Virology, Uppsala University, Uppsala, Sweden*

ANDREW M. BORMAN • *Mycology Reference Laboratory, Health Protection Agency, South-West Regional Laboratory, Bristol, UK*

ORSOLYA BOUQUET • *Department of Analytical and Environmental Chemistry, Faculty of Sciences, University of Pécs, Pécs, Hungary*

MARIA DA LUZ MARTINS • *Instituto de Higiene e Medicina Tropical/Centro de Recursos Microbiológicos, Universidade Nova de Lisboa (IHMT/UNL), Lisboa, Portugal*

SOFIA COSTA-DE-OLIVEIRA • *Department of Microbiology, Faculty of Medicine, Cardiovascular Research Unit, University of Porto, Porto, Portugal*

SYBREN DE HOOG • *CBS-KNAW Fungal Biodiversity Centre, Utrecht, The Netherlands*

RONNIE ERIKSSON • *Department of medical sciences, Clinical virology, Uppsala University, Uppsala, Sweden*

CHRISTIAN FRICKE • *Medizinische Dienste SRO, Langenthal, Switzerland*

STEPHAN FRICKE • *Fraunhofer Institute for Cell Therapy and Immunology (IZI), Leipzig, Germany; Institute of Clinical Immunology, University of Leipzig, Leipzig, Germany; Department of Hematology and Oncology, University of Leipzig, Leipzig, Germany*

NADJA HILGER • *Fraunhofer Institute for Cell Therapy and Immunology (IZI), Leipzig, Germany*

JOÃO INÁCIO • *Instituto de Higiene e Medicina Tropical, Universidade Nova de Lisboa (IHMT/UNL), Lisboa, Portugal*

TAKAO ISHIWATARI • *Department of Surgical Pathology, Toho University School of Medicine, Ota-Ku, Tokyo, Japan*

MAGNUS JOBS • *School of Health and Social Studies, Dalarna University, Falun, Sweden*

ELIZABETH M. JOHNSON • *Mycology Reference Laboratory, Health Protection Agency, South-West Regional Laboratory, Bristol, UK*

FERENC KILÁR • *Institute of Bioanalysis, Faculty of Medicine, University of Pécs, Pécs, Hungary; Department of Analytical and Environmental Chemistry, Faculty of Sciences, University of Pécs, Pécs, Hungary*

BÉLA KOCSIS • *Department of Medical Microbiology and Immunology, Faculty of Medicine, University of Pécs, Pécs, Hungary*

ILDIKÓ KUSTOS • *Department of Microbiology, Alder Hey Children's NHS Foundation Trust, Liverpool, UK; Department of Medical Microbiology and Immunology, University of Pécs, Pécs, Hungary*

KATRINA LACEY • *Molecular Diagnostics Research Group, National University of Ireland, Galway, Ireland*

STEFANO LANDI • *Department of Biology, Genetic Unit, University of Pisa, Pisa, Italy*
ANNA LAU • *Centre for Infectious Diseases and Microbiology, The University of Sydney, Sydney, NSW, Australia*

Westmead Millennium Institute, Westmead Hospital, Westmead, Australia

Department of Laboratory Medicine, Clinical Center, National Institute's of Health, Bethesda, MD, USA

T. LION • *Children's Cancer Research Institute (CCRI), Labdia Labordiagnostik, Vienna, Austria*
JAVIER YUGUEROS MARCOS • *Research and Validation of Biomarkers, bioMérieux SA, Marcy l'Etoile, France*
HARUO NAKAYAMA • *Department of Neurosurgery, Toho University Ohashi Medical Center, Meguro, Tokyo, Japan*
TOCHIGI NAOBUMI • *Department of Surgical Pathology, Toho University School of Medicine, Ota-Ku, Tokyo, Japan*
TETSUO NEMOTO • *Department of Surgical Pathology, Toho University School of Medicine, Ota-Ku, Tokyo, Japan*
LUDWIG NIESSEN • *Weihenstephan Center of Food and Life Sciences, TU-München, Lehrstuhl für Technische Mikrobiologie, Freising, Germany*
CHRISTOPHER OELKRUG • *Fraunhofer Institute for Cell Therapy and Immunology (IZI), Leipzig, Germany*
YOICHIRO OKUBO • *Department of Surgical Pathology, Toho University School of Medicine, Ota-Ku, Tokyo, Japan*
MICHAEL PALMER • *Mycology Reference Laboratory, Health Protection Agency, South-West Regional Laboratory, Bristol, UK*
MICHELANGELO PASCALE • *Institute of Sciences of Food Production (ISPA), National Research Council of Italy (CNR), Bari, Italy*
GIANCARLO PERRONE • *Institute of Sciences of Food Production (ISPA), National Research Council of Italy (CNR), Bari, Italy*
DAVID S. PERLIN • *Public Health Research Institute Center, UMDNJ-NJMS, Newark, NJ, USA*
CIDÁLIA PINA-VAZ • *Department of Microbiology, Faculty of Medicine, Cardiovascular Research Unit, University of Porto, Porto, Portugal*
DAVID H. PINCUS • *R&D Microbiology, bioMérieux, Inc., Hazelwood, MO, USA*
S. PREUNER • *Children's Cancer Research Institute (CCRI), Labdia Labordiagnostik, Vienna, Austria*
ARNE C. RODLOFF • *Institute for Microbiology and Infectious Epidemiology, University of Leipzig, Leipzig, Germany*
ACÁCIO GONÇALVES RODRIGUES • *Department of Microbiology, Faculty of Medicine, Cardiovascular Research Unit, University of Porto, Porto, Portugal*
DAISUKE SASAI • *Department of Surgical Pathology, Toho University School of Medicine, Ota-Ku, Tokyo, Japan*
SONIA SENESI • *Department of Biology, Genetic Unit, University of Pisa, Pisa, Italy*
KAZUTOSHI SHIBUYA • *Department of Surgical Pathology, Toho University School of Medicine, Ota-Ku, Tokyo, Japan*

Department of Dermatology, Peking University First Hospital, Beijing, China

Minoru Shinozaki • *Department of Surgical Pathology, Toho University School of Medicine, Ota-Ku, Tokyo, Japan*
Ana Teresa Silva • *Department of Microbiology, Faculty of Medicine, Cardiovascular Research Unit, University of Porto, Porto, Portugal*
Tania Sorrell • *Centre for Infectious Diseases and Microbiology, The University of Sydney, Sydney, NSW, Australia*

The Westmead Millennium Institute, Westmead Hospital, Westmead, Australia
Keith Stanley • *AusDiagnostics Pty. Ltd., Alexandria, NSW, Australia*
Jiufeng Sun • *Department of Parasitology, The Zhongshan School of Medicine, Sun Yat-Sen University, Guangzhou, China*
Arianna Tavanti • *Department of Biology, Genetic Unit, University of Pisa, Pisa, Italy*
Nina L. Tuite • *Molecular Diagnostics Research Group, National University of Ireland, Galway, Ireland*
Megumi Wakayama • *Department of Surgical Pathology, Toho University School of Medicine, Ota-Ku, Tokyo, Japan*
Francesco Zezza • *Institute of Sciences of Food Production (ISPA), National Research Council of Italy (CNR), Bari, Italy*
Yanan Zhao • *Public Health Research Institute Center, UMDNJ-NJMS, Newark, NJ, USA*

Chapter 1

Overview of Invasive Fungal Infections

Nina L. Tuite and Katrina Lacey

Abstract

The incidence of invasive fungal infections (IFIs) has seen a marked increase in the last two decades. This is especially evident among transplant recipients, patients suffering from AIDS, in addition to those in receipt of immunosuppressive therapy. Worryingly, this increased incidence includes infections caused by opportunistic fungi and emerging fungal infections which are resistant to or certainly less susceptible than others to standard antifungal agents. As a direct response to this phenomenon, there has been a resolute effort over the past several decades to improve early and accurate diagnosis and provide reliable screening protocols thereby promoting the administration of appropriate antifungal therapy for fungal infections. Early diagnosis and treatment with antifungal therapy are vital if a patient is to survive an IFI. Substantial advancements have been made with regard to both the diagnosis and subsequent treatment of an IFI. In parallel, stark changes in the epidemiological profile of these IFIs have similarly occurred, often in direct response the type of antifungal agent being administered. The effects of an IFI can be far reaching, ranging from increased morbidity and mortality to increased length hospital stays and economic burden.

Key words: Invasive fungal infection, Rapid, Early diagnostics, Antifungal drug resistance

1. Introduction

Fungi are eukaryotic organisms, found worldwide in a wide range of habitats from soil and rotting vegetation to extreme environments such as deserts and deep sea sediments. Although over 100,000 species of fungi have been described, the Kingdom Fungi is estimated to have in the region of 5 million species (1). Classification was traditionally based on morphological characteristics, such as the size and shape of spores or fruiting structures. Species were also distinguished by their biochemical and physiological characteristics. The advancement of molecular tools, such as DNA sequencing and phylogenetic analysis, has greatly enhanced

Louise O'Connor and Barry Glynn (eds.), *Fungal Diagnostics: Methods and Protocols*, Methods in Molecular Biology, vol. 968, DOI 10.1007/978-1-62703-257-5_1, © Springer Science+Business Media New York 2013

our knowledge of the genetic diversity within various taxonomic groups (2, 3). The existence of fungi is crucial to the survival of many organisms with which they form synergistic associations. They can be predators of invertebrates, pathogens of plants and animals and are also of profound importance to man (1). Fungal plant pathogens can destroy crops. For example, in the USA an estimated $200 million is lost due to fungal crop damage annually despite a $600 million yearly spend on fungicides (4). Systemic fungal infections in pets and other animals represent a huge problem for the veterinary clinician, and despite the fact that no precise numbers of incidence are available it is believed that the increase in fungal infections seen in human medicine is actually mirrored in veterinary medicine (5). Fungi are routinely used in basic research as experimental model systems for investigation of animal cell functions, e.g., *Saccharomyces cerevisiae.* Many fungi are manipulated in the food and pharmaceutical industries as major producers of materials such as beer, bread, wine, citric acid and other food additives, and important medicines such as antibiotics (6). For example, *Penicillium chrysogenum* is the primary commercial source of penicillin. Penicillin is the precursor for most ß-lactam antibiotics, which represents one-third of the antibiotic market and $8 billion in annual sales, with annual worldwide production of penicillin estimated to be over 40,000 metric tons (4).

It has become apparent over the decades that a number of fungal species previously considered to be innocuous environmental inhabitants are in fact capable of causing devastating disease in humans. These organisms can be difficult to identify with current diagnostic methods and also have been found to vary greatly in their susceptibility to antifungal agents. This causes major treatment management problems for the clinician (7). The earliest known record of a fungal infection was a mycetoma of the foot in the Indian Atharva Veda (c. 2000–1000 BC), later described in 1714 by the French missionary Ponticharry as "padavalmika" (foot ant hill) (8). There are four recognized types of fungal infections: superficial skin infections (e.g., athlete's foot, nail infections, and ringworm), superficial mucosal infections (e.g., oral and vaginal thrush), allergic infections (e.g., asthma and chronic sinusitis), and invasive infections (e.g., aspergillosis and fungal pneumonia) (9, 10). Invasive infections are further divided into two, namely, endemic mycoses (usually causing pulmonary disease in otherwise healthy individuals, such as pneumonia) and opportunistic mycoses (the main focus of this chapter, usually nosocomial affecting immunocompromised patients). Examples of endemic mycoses include infection with *Histoplasma capsultum*, *Cocciodiodes immitis*, and *Blastomyces dermatitidis* (11). The importance of diagnosis of endemic fungi is often overlooked and testing is not performed until the patient fails to recover following antibacterial treatment (12).

1.1. Challenges with Fungal Nomenclature

Frequent name changes at all levels of fungal classification, from species level upwards, have been the source of much confusion over the decades. Not surprisingly, this can hinder the accuracy of diagnostics and reporting of species isolated from clinical settings. For example, *Torulopsis* and *Monilia* are both obsolete synonyms for the genus *Candida*. The species *Candida famata* is also known as *Debaryomyces hansenii* and to a lesser extent, *Torulopsis candida* (13–15). Frequently a fungus can have multiple scientific names depending on its lifecycle and sexual state. The sexual form of a fungus is called the teleomorph, and the asexual form is known as the anamorph. Often these forms are physically distinct and have different names, for example, *Candida kefyr* is the anamorph and *Kluyveromyces marxianus* is the teleomorph of this species. An example of a filamentous fungus is *Aspergillus glaucus*, where the teleomorph is called *Eurotium herbariorum* (14). Efforts among researchers are underway to establish the usage of a unified nomenclature in the field of mycology (2, 3). A comprehensive list of medically important fungi, including their synonyms was compiled and published in 1995, and subsequently updated in 1999 for the Journal of Clinically Infectious Disease by McGinnis et al. (13, 14). Web sites such as Index Forum and ITIS compile lists of current names of fungi with reference to older synonyms. A large-scale collaborative project which involved the efforts of many mycologists and taxonomists was published in 2007 (2).

1.2. Challenges with Fungal Diversity and Diagnosis

Recent advances in molecular biology have hugely enhanced our knowledge of fungal diversity and revolutionized fungal taxonomy. Molecular advances, including genome sequencing, provide crucial information on host/pathogen interaction, how the organism reproduces, and how it can persist both in the environment and within the host. It also provides sequence information which can be used to design unique species specific DNA probes for use in new detection systems (4). Molecular biology has lead to different classification systems within the field of mycology. Many challenges have thus been presented pertaining to the emergence of new species, previously thought to be related or part of another species or group. From a diagnostic point of view this presents challenges to the researcher when new species are identified and require differentiation from others. In an attempt to address these classification issues, a comprehensive phylogenetic classification of Fungi has recently been proposed. This was the result of a large multicentre collaborative effort which shows the importance and need for one unified classification system (2).

1.3. Incidence of Opportunistic Fungal Infections

Recent advances in the health care sector have meant that fungal infections are on the increase due in no small part to the expansion of at-risk populations, in addition to the treatment strategies that often result in longer survival rates of these patients (16).

Patients most at risk of opportunistic colonization from an invasive fungal infection (IFI) are those who have undergone solid organ transplant, those in receipt of immunosuppressive or chemotherapeutic agents, those suffering from HIV, pediatric, and elderly patients, and finally, those patients undergoing surgery (17). While bacterial infections are far more prevalent than fungal infections, mortality rates appear to be far higher for fungal infections when compared to bacterial infections (18–20). Filamentous fungi, yeast-like fungi, and dematiaceous fungi are well-known causative agents of IFIs (21); however, *Candida* and *Aspergillus* still remain the most frequently isolated species. The majority of diagnostic tools have been developed with these two organisms in mind (22) as together, *Candida* and *Aspergillus* account for approximately 90% of all nosocomial fungal infections (11). Morbidity–mortality rates for high risk patients suffering from either of these two pathogen groups are in the region of 40–50% for *Candida* while for *Aspergillus* the numbers stand between 80 and 100% (23).

2. Clinically Relevant Fungi

2.1. Medically Important Yeasts

Hippocrates was the first to describe a yeast infection when he wrote of thrush in the fifth century BC (8). Detection of yeast cells in thrush by microscopic techniques did not occur until 1839. It was thought that *Candida albicans* was the only yeast species capable of causing human infection. As advances in the medical field ensued it soon became apparent that other yeast species were also clinically relevant. Despite this, little medical concern was given to them until relatively recently (7). The advent of new cancer treatment regimes, increased use of intravenous catheters, and other medical developments essentially prolonging the lives of immunocompromised patients has resulted in a major shift in the epidemiology of yeast infections and since the 1960s there has been a steady rise in the number of opportunistic yeast species causing severe human infections (7).

2.1.1. Candida Species

Candida species are normal commensals of the skin, mucosa membranes, and gastrointestinal tract of humans and other mammals and are responsible for most nosocomial fungal infections. Of the approximate 100 known species of *Candida*, relatively few (12–14) have been associated with human infection (11, 24). *Candida* species can cause a wide spectrum of disease, from superficial infections such as thrush and nail bed infections to serious life-threatening illnesses such as endocarditis, meningitis, osteomyelitis, and candidemia (11). In the USA, candidemia is the fourth most common bloodstream infection (25).

The route of infection for *Candida* species is typically via intravascular devices or through the gastrointestinal tract (26). Interestingly, one study found that about 70% of hospital personnel harbored yeasts on their hands (27). While there is no doubt that *Candida albicans* is the most frequently isolated *Candida* species from clinical specimens (24), when combined with other *Candida* species, or "*Non-albicans candida*" (NAC) such as *C. glabrata*, *C. parapsilosis*, *C. tropicalis*, and *C. krusei*, together they account for roughly 99% of all human episodes (28). This is of particular concern since many NAC have intrinsic resistance to current antifungal agents, added to this are the technical difficulties associated with diagnosis as it is thought best to diagnose to species level (11, 29, 30). A retrospective surveillance study conducted in 2000 found that the species responsible for *Candida* bloodstream infections were as follows: *C. albicans* (54%), *C. glabrata* (16%), *C. parapsilosis* (15%), *C. tropicalis* (8%), *C. krusei* (1.6%), and other *Candida* species (4.6%) (25, 30). *C. tropicalis* has been reported to be a common cause of fungemia where risk factors are similar to that of *C. albicans* (24). *C. krusei* has been reported to spread from the gastrointestinal tract of severely immunocompromised patients to cause fungemia and endophthalmitis and is of particular concern because of its intrinsic resistance to some antifungal drugs (11). Although these aforementioned species are the most common NAC species, *C. lusitaniae* and *C. guilliermondii* are emerging as important species causing IFIs (11, 24, 31). *C. rugosa* is a common veterinary pathogen however it has been isolated from wounds of patients in burn units and from both blood and urine of hospitalized patients (11, 32). A number of other NAC species such as *C. famata*, *C. cifferri*, *C. pulcherrima*, *C. utilis*, *C. catenulata*, *C. norvegiensis* have been isolated from clinical samples and account for <1% of candidemia (31).

Research into diagnostics for detection of *Candida* infection has largely focused on systemic disease although methods for diagnosis of vaginal candidiasis have also been evaluated. It is thought that as the use of molecular methods of detection become more widespread that the cost of such tests will reduce over time, increasing their use for diagnosis of superficial infections (33). The diagnosis of invasive *Candida* infections is challenging, symptoms can be nonspecific, and positive cultures may not be obtained until late in the infection (29, 34). The Infectious Disease Society of America recently published guidelines for the treatment of *Candida* infections (35).

2.1.2. Cryptococcus Species

Cryptococcus neoformans is an environmental saprophyte and has a worldwide distribution. An opportunistic human pathogen, it is the most common etiological agent of cryptococcosis. It is found in the environment and clinical setting as budding yeast. Cells are spherical and are protected by a polysaccharide capsule. This polysaccharide is the diagnostic target for the cryptococcal antigen test. *C. neoformans var gattii* also known as *Cryptococcus gattii* can also cause infection and is found primarily in tropical regions. *C. neoformans* can cause infection in any organ but predominantly infects the CNS and lungs. Pulmonary cryptococcosis, most commonly caused by inhalation of *Cryptococcus* cells, may be asymptomatic or may present with nonspecific symptoms including cough and fever. Cryptococcal meningitis has become widespread in recent times, specifically with the spread of HIV and the use of immunosuppressive drugs. It has become a common opportunistic infection among late stage AIDS patients and solid organ transplant patients and other immunocompromised hosts but also is often reported in individuals who seem to be otherwise immunocompetent. Because of the nonspecific nature of these symptoms, cryptococcal infection should be tested for should symptoms of meningitis occur in the relevant clinical setting. Diagnosis of cryptococcal infection can be through microscopy using Indian ink stain, positive blood or CSF cultures, serology, or histology. Diagnosis is rarely difficult in HIV-associated cryptococcal infection due to the high organism load. But in non-HIV-associated infection diagnosis can be difficult due to false-negative cultures and antigens can result from low organism yield. CT scans, large volume CSF cultures, and lumbar punctures may be needed (36). Rapid and accurate diagnosis is crucial since untreated cryptococcal meningitis is fatal. Treatment is aggressive, usually using a combination of fluconazole and amphotericin B. Unlike *Candida* and *Aspergillus*, enchinocandins have little activity against *Cryptococcus neoformans* (36).

2.1.3. New and Emerging Yeast Pathogens

Since yeast infections are not notifiable diseases, there is no central database available to record isolations of specific yeast species from year to year. As a result of this, it is thought that the actual number or incidence of such diseases is under reported. It has been suggested that case reports in the literature significantly under represent the incidence of infection with emerging yeasts. This is thought to be due to investigators reluctance to publish non-novel data. Nevertheless, organisms emerging as important yeast pathogens include *Malassezia*, *Rhodorula*, *Hansenula*, and *Trichosporon* species. However, this spectrum of organisms is growing (7). *Trichosporon* species normally cause a superficial infection of the hair shaft called white piedra; however, recent reviews have described manifestations of trichosporonosis including severe skin infections, endocarditis, peritonitis, and bloodstream infections. *Malassesia* species are frequently being observed as nosocomial pathogens (11).

3. Medically Important Filamentous Fungi

A number of fungi previously thought to be non-pathogenic, including the *zygomycetes* and haline and dematiaceous molds, are capable of causing opportunistic infections in humans. Although *Aspergillus* species account for most cases of invasive mold infections, a number of other species are being recognized as important pathogens causing devastating and often fatal diseases (17).

3.1. Aspergillus Species

Aspergillus species are ubiquitous in nature, commonly occurring in soil, water, and vegetation. *Aspergillus* species are opportunistic human pathogens and are the most common clinically associated invasive molds, primarily *A. fumigatus*, although *A. flavus*, *A. niger*, and *A. terreus* are increasingly being recognized as important human pathogens (34, 37). Aspergillosis was first described in the 1940s, since then it has become a major problem and is now the leading cause of death among IFIs (38). This enormous increase in invasive aspergillosis (IA) cases has been linked to the ever increasing immunocompromised host population which has resulted from immunosuppressant therapies and advances in medical procedures, previously unheard of. Risk factors associated with aspergillosis include prolonged granulocytopenia, development of graft-versus-host disease, immunosuppressive therapy, use of adrenal corticosteroids, diseases such as chronic granulomatous disease, AIDS, cancer, solid organ, and bone marrow transplants (39). *Aspergillus* species can cause invasive aspergillosis, aspergilloma, chronic necrotising aspergillosis, tracheobronchotic aspergillosis, but colonization without infection can also occur (34). Invasive aspergillosis is associated with a mortality rate of approximately 85%. *A. fumigatus* is the most common species to cause invasive aspergillosis, causing in the region of 90% of cases worldwide (37, 40). The primary route of *Aspergillus* infection is thought to be through inhalation of conidia. Dissemination to other organs is thought to occur following invasion of the lung tissue. Exposure of the immunocompromised host to environmental isolates plays a role in the pathogenesis of this disease (39). These fungi have been isolated in hospitals from air ventilation systems, carpets, and dust dislodged during construction (41). Stringent management and control measures should be paramount. Another proposed route of *Aspergillus* infection is through ingestion of contaminated food, although no outbreak of *Aspergillus* infection has been reported. To date some localized infections have been associated with contaminated wound dressing or tape (11). Like *Candida* species, the most useful way to select an appropriate treatment regime comes from identification to the species level (30). A suitable rapid diagnostic test for *Aspergillosis* is severely lacking. At present, proof of IA infection can only be shown by growth in tissue or culturing of the fungus from the test specimen, and in many patients, proof is only found at autopsy (42). Diagnosis is currently via CT scan, open lung biopsy, microscopy, and culture (34). Early treatment of invasive aspergillosis is essential. Primary treatment involves Amphotericin B; however, surgery may be necessary for localized infections (37, 43).

4. Emerging Molds

The incidence of infection involving *Fusarium* species has dramatically increased in recent years (44–46). *F. solani* is the predominant species isolated (44). The only antifungal agent effective against *Fusarium* is amphotericin B, unfortunately few patients survive disseminated *Fusarium* infections despite treatment (39). Other opportunistic filamentous fungi often involved in cases of fungemia, disseminated infections, and fungal pneumonia include: *Acremonium* species, *Scedosporium* species, the class *Zygomycetes*, *Paecilomyces* species. Some of these mimic the clinical symptoms of aspergillosis and thus accurate diagnosis is essential (39). *Scedosporium* species are emerging as human pathogens among immunocompromised hosts, in particular, *S. apiospermum* and *S. prolificans.* Such infections are usually difficult to treat because of resistance to current antifungal therapies and mortality rates are extremely high (46). *Zygomycetes* are a class of fungi typically found in soils and rotting vegetation and are increasingly seen as opportunistic human pathogens. Routes of infection can be ingestion, inhalation, or through percutaneous inoculation of spores (47). *Zygomycetes* have emerged as important pathogens in immunocompromised patients particularly those belonging to the order *Mucorales* (e.g., *Mucor* species and *Rhizopus* species). Another order of *Zygomycetes*, the *Entomphthorales*, which are principally insect pathogens, are now frequently implicated in human disease, e.g., *Conidiobolus* and *Basidiobolus* species (48). The increase in incidence of human infection with *Zygomycetes* is thought to be linked to more widespread use of the antifungal drug, voriconazole which has no activity in vitro against the *Zygomycetes* (46, 48). Mortality rates are as high as 80% (47). Successful management relies on early detection. Treatment involves urgent surgical debridement of infected tissue and initiation of suitable antifungal therapy (46–48).

5. Diagnosis of IFIs

Proven, probable, and possible are the three levels of classification used to identify the existence of an IFI. The term proven can be applied to all types of patient, be they immunocompromised or not, however, both probable and possible refer to those patients who have been categorized as being immunocompromised (49). The introduction and acceptance of such standard terminology in defining types of IFI is of critical importance, adding credence to the consistency and reproducibility of clinical studies. Not withstanding the importance of correct classification of an IFI the fact still remains that early diagnosis is vital, and therefore it is imperative that treatment is started as soon as an infection is suspected as any delay in treatment leads to increased morbidity and mortality (50).

5.1. Traditional Methods of Diagnosis

Early and accurate diagnosis of life threatening fungal infections is of paramount importance to allow timely initiation of antifungal therapy and to reduce mortality rates. A clinician faces a myriad of challenges when attempting to accurately diagnose and treat an IFI. Historically, the detection and identification of fungi has depended largely on the more traditional based methods such as histology, microscopy, and culture-based techniques. Although considered the cornerstone of proving the presence/absence of a fungal disease, their diagnostic worth is limited. Despite recent advances in diagnostic methods, microscopic examination and cultivation of clinical samples are still considered to be the "gold standard" method of identification. However these methods are not without limitations, they can result in low sensitivity and specificity, and often only give positive results in the later stages of infection (49).

5.1.1. Culture

Culture-based diagnostic methods are fraught with difficulties, for example, in order to obtain a biopsy to be used for either culture or histopathology from a sterile site such as the lung, an already critically ill patient would need to be subjected to an invasive technique. In addition, when a sample has been proven to be positive, doubt remains over whether the result of a biopsy taken from a non-sterile site is actually down to colonization or due to an active infection (51). Due to its transient nature, blood cultures have been shown to have only a 50% and a 10% success rate for diagnosing candidemia and IA, respectively (52). However, due to the longer incubation times required for growth before starting specific treatment, survival is severely impacted. Similarly, when a sputum or bronchoalveolar lavage (BAL) is taken from a patient with suspected invasive aspergillosis (IA), frequently the organism cannot be cultured (53). One of the most important roles of the clinician is to determine if a patient requires antifungal treatment in the early period of infection. Therefore it is of paramount importance that a rapid test for early diagnosis is developed (22). Frequently, primary identification media such as CHROMagar Candida media for culturing *Candida* can be used for the initial presumptive identification of some of the most medically important species of *Candida* such as *C. albicans*, *C. glabrata*, *C. tropicalis*, and *C. parapsilosis*. In addition, the majority of yeasts arising from clinical samples can be identified using commercially available biochemical kits such as API 20C AUX, VITEK 2, and RapID Yeast Plus (54). Despite the user friendly nature of these kits, the identification process does not end here, instead, a suite of additional morphologic based tests must be performed to avoid misidentification of those microorganisms showing identical biochemical profiles (53).

5.1.2. Histopathology

Histopathology is an important diagnostic tool and refers to the microscopic examination of infected tissue. It is a rapid and cost effective method for diagnosing the presumptive or definite presence of an IFI (17). While it is not always possible to retrieve a biopsy from a critically ill patient, this method is further compounded by the fact that several different organisms can display similar histopathological profiles, thereby rendering it almost impossible to identify a specific pathogen based on morphological traits alone. For example, many *Fusarium* sp., *Pseudallescheria* sp., and *Penicillium* are all alike in that they share hyaline and narrow septate hyphae making them indistinguishable from *Aspergillus* in tissue biopsies (55, 56). This inability to definitively identify a specific organism can severely affect the outcome for a patient (57).

5.2. Fungal Cell Wall Markers

When a clinician suspects the presence of an IFI often the patient is treated empirically with antifungal therapies that may include the unnecessary administration of potentially harmful and costly drugs (58). Due to the potentially toxic nature of this approach, interest has increased in the use of what is known as pre-emptive antifungal therapy which refers to the deferred treatment until there is sufficient evidence to confirm the presence of an IFI (59). Therefore, intense efforts have been put into the development of laboratory markers in an attempt to reduce the diagnosis time of an IFI (53). These markers are known as galactomannan (GM) and 1,3-beta-D-glucan (BG) and mannan. This in turn has led to a shift in the way that IFIs can be prevented and treated more efficiently (60).

5.2.1. Galactomannan

Definitive diagnosis of invasive aspergillosis is of the utmost importance to allow for early initiation of antifungal treatment (22). Galactomannan is a cell wall polysaccharide released by *Aspergillus* species into extracellular fluid during fungal growth in tissue (61). For early diagnosis of IA, GM antigen monitoring has proven its importance as a noninvasive diagnostic tool (62). Before clinical manifestation of IA, circulating GM can be detected anywhere between 5 and 8 days before fungal burden is obvious. Not only is GM detectable in serum or plasma (63) but it can also be detected in BAL (64) and cerebrospinal fluid (65) using a sandwich type enzyme linked immunosorbent assay (53). Following analysis it appears that the BAL fluid assay is more sensitive than the serum assay (66). However, as with other methods of diagnosis this assay is not without its limitations, resulting in contradictory results due to a number of factors including: prior treatment with antifungal therapy affecting the levels of circulating fungal components (67), false-positive results in conjunction with use of antibiotic treatments (68), and finally, the range of cut offs of positivity across different studies (60). According to a recent meta-analysis of serum galactomannan in patients with neutropenia and/or hematologic malignancy sensitivity results of 71% coupled with a specificity of 89% were reported for those definitive cases of invasive aspergillosis (69). It is clear that the diagnosis of invasive aspergillosis should not be based on a single test alone and that the BAL fluid galactomannan assay should be used as an adjunct to further tests.

5.2.2. 1,3 Beta-D-Glucan

1,3 Beta-D-glucan is a component of the cell wall of most fungi which can be detected in the blood during an IFI. There are two notable exceptions, namely, the *Zygomycetes* which do not produce BG and *Cryptococcus* species which release such low levels of BG that it cannot be detected in human serum (70). The test is of significance due to its ability to detect infections caused by such species as *Fusarium*, *Trichosporan*, *Saccharomyces*, and *Acremonium*. While these species are undoubtedly not as common as *Aspergillus* and *Candida*, they are extremely dangerous organisms for the immunocompromised patient (71). Limited data exists as to the efficacy of this test as a diagnostic tool; however, it may useful as an early identifier of infection and is reputed to be highly sensitive (72). Of the commercially available assays, two in particular are most commonly used, namely, Fungitec-G and Fungitell, manufactured by Sikagaku Kogyo Corporation and Associates of Cape Cod, respectively. A possible pitfall of the Fungitec-G assay is that, in the absence of an IFI, medical sources of BG can lead to a positive test. For example, filters and dialysis membranes made from cellulose contain BG, as do cotton gauzes and sponges used in surgery, and some drugs. Unfortunately only a limited number of studies have been performed to date so literature is limited. Those studies that are available have reported sensitivities in the range of 70–100% while specificity ranges anywhere from 76 to 83.8%. Again, these results must be viewed with caution, as only a limited number of invasive aspergillosis cases have been subjected to testing with this assay (72, 73). Thus, further more intensive testing must be performed before this assay is put forward as a viable diagnostic option (74).

5.2.3. Mannan

Mannan is the major *Candida* cell wall antigen and is the substrate for one of the most extensively studied antigen tests for detection of systemic candidiasis which was proposed as far back as 1979 (75). A vast quantity of literature has amassed over the years, all of the similar opinion that a positive mannan test may actually correlate with invasive candidiasis. As detection of the infection at an early stage is critical, the immunoenzymatic Platelia Candida Ag (Bio-Rad) test is performed in parallel with the anti-mannan antibody test (Platelia Candida Ab/Ac/Ak; Bio-Rad) (76). Mannan occurs at low levels and is rapidly cleared from the bloodstream, therefore, one disadvantage of using this marker is the frequency with which tests must be performed and thus the increased cost (77).

5.2.4. Imaging

Chest computed tomograph (CT) scans have proven to be a very useful tool, especially in the early stages of infection (78). More specifically, the "halo sign," which is known as a region of ground-glass attenuation surrounding a pulmonary nodule on CT scan of the chest, is considered to be an early indicator of invasive pulmonary aspergillosis (IPA) (79). However, using the "halo sign" for diagnosis is not without its difficulties, the reason being that the CT scan must be performed within 5 days of the onset of a suspected infection, if not, then approximately 75% of the halo signs disappear within a week (80). The diagnostic use of the subsequent "air crescent" sign is limited by the fact that it only becomes visible in the third week of infection which may be too late to begin treatment of an invasive aspergillosis infection (81). To conclude it is clear that the CT halo sign has more diagnostic potential than the later air crescent sign, as treatment can be started immediately.

5.2.5. Matrix Assisted Laser Desorption/Ionization Time of Flight Mass Spectrometry

In recent years, MALDI-TOF MS has been introduced to the clinical laboratory for rapid species identification (82, 83). The method works by analyzing the mass patterns from crude cell extracts of an isolate and comparing to a database of patterns of reference strains. A recent study evaluated the potential use of two commercially available MALDI-TOF MS systems for their application in clinical diagnostics using over 1,000 yeast and yeast-like clinical isolates from geographically distinct locations. The investigators compared the performance of both kits to classical clinical identification methods using microscopy and biochemical techniques (82). It was reported that the identification of pathogenic yeasts using MALDI-TOF MS in the clinical laboratory will greatly improve fungal diagnostics and hence improve treatment regimes. One of the main findings was that MS could be performed in a fraction of the time it takes for classical techniques. Another important advantage found was that MS could differentiate closely related yeast species. For example, both methods could differentiate the *Candida* ortho meta parapsilosis cluster, whereas the classical methods could not. This holds significant clinical importance since different antifungal susceptibility profiles have been observed within this cluster (82). Another recent article describes a MALDI-TOF MS method for precise and rapid identification of common clinically isolated *Aspergillus* species as well as newly reported species in the clinical setting (83). The authors suggest that since the procedure only takes a few minutes, introduction of these techniques into main stream clinical practice will not only provide a more rapid diagnostic process but also lead to more accurate identification of fungal species (82, 83).

5.3. Molecular Diagnostics

In theory, an ideal molecular test should be capable of providing absolute sensitivity without adversely affecting specificity and should have the ability to rule infection in or out. However, this is rarely the case, if assays are extremely sensitive then inevitably they will generate false-positive results, and a compromise is needed that will find a balance between the early detection of subclinical infections and at the same time between the early detection of subclinical infections and low level contamination (84). Timing is critical for a patient suffering from an IFI. Due to the changing epidemiology and increase in emerging fungal pathogens, there is a knock on affect, causing an increased demand for more broad-spectrum diagnostic tests. The overall aim of using molecular methods for diagnostics is the hope that they will provide superior specificity, sensitivity, and turn around time. Many published accounts are available, which outline the use of molecular methods for diagnosis, most specifically real-time PCR. While these assays show great promise, being sensitive and specific, their usefulness is limited to single generas, for example, *Candida* (85, 86) or *Aspergillus* (87). In order to address this problem, Landlinger et al. developed a panfungal real-time PCR assay based on the 28S ribosomal RNA multicopy gene, which facilitated the detection of at least 80 pathogenic species (88). Despite its obvious value, PCR assays have not been widely accepted and as such have been side tracked by other diagnostic methods such as antigen detection assays. However as previously discussed, these methods are far from perfect and often fail to detect all of the infection causing fungal pathogens. Nonetheless, PCR assays still offer several advantages over other methods such as imaging, culture, and histopathology. Due to the nature of PCR, it has the supreme ability to detect minute amounts of starting material, especially when targeting a specific gene that has multiple copies (89). In addition, PCR assays have a multitude of options when it comes to the design process, ranging from complete genera down to a single species.

Not only can a PCR assay be designed to be specific but it can also be quantitative, allowing the clinician to determine the fungal load of the particular infection and the ability to ascertain how far the infection has progressed. While there is no denying the importance of molecular methods for fungal diagnostics and its potential for improving patient survival, it is still somewhat thwarted by a lack of standardization (90). This, coupled with the lack of a commercial system that has been rigorously tested means that routine PCR testing is not incorporated into the European Organization for Research and Treatment of Cancer/Mycoses Study Group (EORTC/MSG) criteria (91). Again, as seen with all other available diagnostic approaches for IFIs, results when using PCR vary, with sensitivities reported as being anywhere from 45 to 92%, with sensitivity being high at greater than 90% (85, 92, 93). Most notably is an analysis performed by Mengoli et al. which assessed the efficacy of the PCR assays in diagnosing IA (94). Sample types included plasma, serum, and blood. Analysis revealed that sensitivity and specificity were 75% and 87%, respectively, for two positive samples, while for a single positive sample the results were 88% and 75%, respectively. Clearly the difference in these percentages was of concern, and obviously a single negative PCR result cannot be considered to exclude the presence of IA. It seems most likely that future diagnostic efforts, for the reliable detection of IFIs, will rely on a combination of diagnostic procedures as opposed to the sole reliance on any single method. This approach will promote more informed decisions on treatment strategies that can be implemented by clinicians.

6. Treatment of IFIs

According to the National Nosocomial Infection Surveillance Program, there has been a decrease in hospital bed numbers in the USA, while conversely, ICU has seen bed numbers increase (95). This increase is attributable to the improvement in supportive medical care in the ICU. This improvement may be viewed as a double edged sword since critically ill patients have prolonged hospital stays and improved survival rates and are as such vulnerable to IFIs (96).

Broadly speaking, there are four main approaches to the timely administration of antifungal agents for the management of a suspected IFI, namely, prophylactic, empiric, pre-emptive, and finally target therapy. The main difference between all four treatment options is the timing, with prophylaxis being the earliest and target being the latest to be given. Prophylactic therapy refers to the administration of antifungals to patients who are considered to be at risk of developing an IFI but have shown no attributable signs or symptoms, the main goal being prevention (97). Notwithstanding the obvious benefits of this type of therapy, there are however a number of shortcomings including the high cost of treatment not to mention the associated risks of giving drugs to individuals before an IFI has been confirmed or indeed excluded. Those patients deemed to be at a high risk of developing an IFI and have signs and symptoms indicative of infection are given empiric therapy, as with prophylactic therapy an actual organism has not been identified (98, 99). Pre-emptive therapy is that given to a patient who has had early diagnostic tests performed, and evidence suggests that an IFI is likely. Finally, targeted therapy as the name suggests is given once an actual pathogen has been identified by histopathology and/or culture (97–99) and therefore treatment can be very specific. As with all other treatment options these four approaches are not without their advantages and associated disadvantages (100).

6.1. Antifungal Therapy

For some time Amphotericin B, first generation azoles, and flucytosine were the only available treatment for fungal infections. The emergence of antifungal resistance isolates and also the toxic effects observed with some of these agents prompted the development and introduction of new formulations and classes of antifungal drugs. This has greatly improved treatment options available to clinicians. Essentially only four classes of established antifungal drugs are available, namely, polyenes (e.g., amphotericin B), azoles (e.g., fluconazole and itraconazole), allylamines (e.g., terbinafine), and the newly introduced echinocandins (e.g., caspofungin). Of these aforementioned classes, only three are used to treat systemic fungal infections (10). Although the development of new antifungal agents has significantly contributed to the successful treatment of fungal diseases, their effectiveness depends on the fundamental understanding of how these drugs interact with concomitant medications in addition to their associated toxicity. Clearly, for management of IFIs and treatment with such antifungal compounds, an in-depth knowledge of their pharmacokinetic and pharmacodynamic properties is essential (101). In addition, it is a well-known fact that inappropriate antifungal use actually adds to the global increase in antifungal resistance and may in fact lead to a variety of adverse outcomes, including unnecessary exposure to antifungal drugs, continual infections, and an associated increase in hospital costs (102).

6.1.1. Polyenes

Polyenes are by far the oldest category of antifungal agents (103) and while in excess of 200 polyene antibiotics have been identified, amphotericin B and nystatin are the only polyenes that are routinely used in a clinical setting. They have been in use since the 1950s and up until the mid-1980s amphotericin B was the gold standard of antifungal therapy (104), but, alternatives were scarce so it was hard to gauge just how effective it was. Despite its high in vitro activity against a broad spectrum of pathogens, an important drawback of this antifungal agent lay in the fact that it causes acute toxicity negatively affecting the kidneys (105). In order to moderate the effects of drug toxicity seen when using amphotericin B, newer lipid preparations of amphotericin B, including amphotericin B lipid complex (Abelcet; Enzon), liposomal amphotericin B (Am-Bisome; Astellas Pharma US), and amphotericin B colloidaldispersion (Amphotec; Three Rivers Pharmaceuticals) were developed (106). Despite the fact that administering these newer formulations caused a reduction in the percentage of renal toxicity, none of them proved themselves to be more superior to the original molecule, and all versions still have to be given intravenously.

6.1.2. Azoles

Azoles are effective against a suite of fungal pathogens and act by inhibiting the synthesis of ergosterol of the fungal cell membrane. In the 1980s the first azole antifungals, the imidazoles miconazole and ketoconazole, appeared in the USA (103). In addition, two newer broad-spectrum triazoles (voriconazole and posaconazole) have been added to the collection of antifungal agents available to the clinician in order to combat serious fungal infections (107–110).

6.1.3. Echinocandins

Echinocandins (caspofungin, micafungin, and anidulafungin) are the currently approved agents for clinical use (111) and are a new class of parenteral antifungal agents that target the fungal cell wall (29). Although originally discovered in the 1970s, the echinocandins are the most recent class of antifungal agents to be introduced with caspofungin being approved for use in the USA in 2001. The echinocandins are active against *Aspergillus* and *Candida* species, including azole resistant *Candida* species. They are fungicidal in nature causing rapid lysis in growing cells and have shown fewer drug to drug interactions (112). These three agents are pharmacologically similar and only differ in a number of traits (29). Almost all *Candida* show in vitro susceptibility to the echinocandins and by virtue of this they have been approved for the treatment of candidemia and other forms of invasive candidiasis (113). In case of infection with *C. glabrata* or *C. krusei*, echinocandins are preferred over azoles. One disadvantage associated with the use of echinocandins is the fact that they have insufficient bioavailability for oral use and thus must be administered intravenously. In sum, the echinocandins remain the treatment of choice when it comes to *Candida* species; however, their role in the treatment of invasive aspergillosis remains an unknown quantity.

6.2. Alternative Treatment Strategies to Antifungal Treatment

In addition to the dispensing of antifungal agents, urgent debridement of infected tissue, where possible, can significantly improve the patients' chance of survival, particularly in the case of filamentous fungal infections (46, 48, 114). In addition, it is also recommended that intravenous lines and catheters which may be the source of infection be removed (46, 114). Minimizing or reversing immunosuppression and neutropenia by reducing the use of immunosuppressive drugs, e.g., steroids, is thought to improve the patients outcome. Some studies have suggested that the use of immunostimulatory drugs (e.g., granulocyte transfusions, cytokines (such as G-CSF, GM-CSF, and INF-γ) in combination with antifungal therapy) is useful and enhances the patients chance of recovery (46, 115, 116).

7. Antifungal Resistance

The increased use of antifungal drugs in recent years has lead to the development of antifungal resistance amongst fungal clinical isolates. This is a worrying concern and has been the focus of much research in recent times. There are various aspects of antifungal resistance which many researchers have recently focused on, understanding the mechanisms of antifungal resistance, alternative treatment options for infections caused by resistant organisms, methods to detect resistance, and strategies to prevent and control the spread of resistance (117). Antifungal resistance can be either intrinsic (i.e., "natural" resistance) or acquired (develops following exposure to the antifungal agent in question). A third type of resistance has also been described referred to as clinical resistance. This type of resistance is observed in severely immunocompromised patients with consistent relapse of infection with an isolate which when tested in vitro appears to be susceptible to the antifungal agent. It is thought that a possible explanation for clinical resistance is due to suboptimal concentrations of the antifungal agent in blood due to interference of other drugs used to treat the patients condition (30, 118). Mechanisms of resistance can include overexpression of efflux pumps to essentially prevent intracellular accumulation of the antifungal drug and genetic modification or overexpression of the antifungal targets. Understanding the specific resistance mechanisms is crucial to finding alternative treatment drugs for resistant strains.

The most prevalent antifungal resistance is that of *Candida* species to azoles (119). Accurate identification of *Candida* to species level is invaluable since it can infer the likely antifungal susceptibility and therefore helps in the selection of a suitable antifungal treatment. For example, *C. glabrata*, *C. krusei*, and *C. rugosa* frequently show resistance or reduced susceptibility to fluconazole, whereas *C. albicans*, *C. parapsilosis*, and *C. tropicalis* are reliably susceptible (32, 46). Identification to species level is also recommended for *Aspergillus* since various species also exhibit different resistance/susceptibility profiles (30).

There is an ever increasing need for standardized drug susceptibility testing, and a method which is fast, accurate, reproducible, and inexpensive is desired. Standard microdilution susceptibility test methods have been developed by the Clinical Laboratory Standards Institute, (CLSI, formerly the NCCLS) for yeasts (*Candida* and *Cryptococcus* species; NCCLS, M27-A2) and molds (*Aspergillus*, *Fusarium*, *Rhizopus*, *Pseudallescheria*, and *Sporothrix* species; NCCLS, M38-A). A standard antifungal disk diffusion susceptibility testing method was also developed for *Candida* species (CLSI, M44-A). European Committee for Antimicrobial Susceptibility Testing (EUCAST) subsequently developed a broth dilution test for susceptibility of yeast species (120). Antifungal susceptibility testing should be routinely performed as this can provide valuable clinical information and aid greatly in treatment decisions, particularly as new resistance isolates come to light (46). In order to prevent and control antifungal resistance, a number of measures should be introduced, including the prudent use of antifungal agents, avoidance of low dose therapy, encouraged use of combination therapy, diagnosis of the etiological agent, treatment with the appropriate antifungal drug, and finally, regular testing and surveillance of antifungal resistance (117, 118).

8. Conclusions and Future Perspectives

IFIs will continue to be a major challenge for the clinical sector, and regrettably, the optimal approach to diagnosing IFIs still remains uncertain. Timing appears to be the single most critical factor if a patient is to survive the ravages of such an infection. Traditional approaches such as direct microscopy, cultivation, and histopathological evaluation still remain the gold standard for diagnosis of IFIs. Great strides have been taken in an attempt to improve the accuracy and speed of diagnosis of these devastating infections, allowing anti-mycotic treatment to begin as early as possible. A number of molecular methods have been developed in recent years; however, have not yet been standardized and have thus far been used only in experimental studies. It has been reported that the most convenient non-culture-based methods for diagnosis of IFI and monitoring antifungal treatment are commercial systems which detect fungal cell wall antigens galactomannan and 1,3 B-glucan (22). DNA and RNA methods appear to hold great promise for improved specificity and sensitivity; however, these methods need to be validated and standardized before they can be employed in routine clinical laboratory testing. Further, an important factor which cannot be ignored when selecting a new diagnostic test for IFIs is the cost. Efforts among researchers continue in order to improve the outcome of fungal disease. Many resources are currently available, for example, the WHO Collaborating center for the Mycoses "An international center of excellence, developing and promoting cost effective strategies for the diagnosis, prevention and control of mycotic diseases." Other useful resources which provide guidelines for diagnosis and treatment of fungal infections include the European Organization for Research and Treatment of Cancer/Invasive Fungal Infections Cooperative Group and the National Institute of Allergy fungus. Other useful resources which provide guidelines for diagnosis and treatment of fungal infections include the European Organization for Research and Treatment of Cancer/Invasive Fungal Infections Cooperative Group and the National Institute of Allergy and Infectious Diseases Mycoses Study Group.

References

1. Blackwell M (2011) The Fungi: 1, 2, 3 … 5.1 million species? Am J Bot 98(3):426–438
2. Hibbett DS et al (2007) A higher-level phylogenetic classification of the Fungi. Mycol Res 111(5):509–547
3. Celio GJ et al (2006) Assembling the fungal tree of life: constructing the structural and biochemical database. Mycologia 98(6):850–859
4. Birren B, Fink G, Lander E. (2004) A white paper for fungal comparative genomics, in fungal genome initiative. Centre for Genome Research, 320 Charles Street, Cambridge
5. Wiebe V, Karriker M (2005) Therapy of systemic fungal infections: a pharmacologic perspective. Clin Tech Small Anim Pract 20(4): 250–257
6. May GS, Adams TH (1997) The importance of fungi to man. Genome Res 7(11):1041–1044
7. Hazen KC (1995) New and emerging yeast pathogens. Clin Microbiol Rev 8(4):462–478
8. Ainsworth GC (1986) The history of medical and veterinary mycology. Cambridge University Press
9. Ascioglu S et al (2002) Defining opportunistic invasive fungal infections in immunocompromised patients with cancer and hematopoietic stem cell transplants: an international consensus. Clin Infect Dis 34(1):7–14
10. Denning DW (2004) Antifungals. Where are we headed? Drug plus international http://pharmalicensing.com/articles/disp/1095693369_414ef439d6909
11. Fridkin SK, Jarvis WR (1996) Epidemiology of nosocomial fungal infections. Clin Microbiol Rev 9(4):499–511
12. Wheat LJ (2009) Approach to the diagnosis of the endemic mycoses. Clin Chest Med 30(2):379–389
13. McGinnis MR, Rinaldi MG (1995) Selected medically important fungi and some common synonyms and obsolete names. Clin Infect Dis 21(2):277–278
14. McGinnis MR, Sigler L, Rinaldi MG (1999) Some medically important fungi and their common synonyms and names of uncertain application. Clin Infect Dis 29(4):728–730
15. Carrasco L et al (2005) Isolation of Candida famata from a patient with acute zonal occult outer retinopathy. J Clin Microbiol 43(2): 635–640
16. Naggie S, Perfect JR (2009) Molds: hyalohyphomycosis, phaeohyphomycosis, and zygomycosis. Clin Chest Med 30(2):337–353, vii–viii
17. Guarner J, Brandt ME (2011) Histopathologic diagnosis of fungal infections in the 21st century. Clin Microbiol Rev 24(2):247–280
18. Castagnola E et al (2008) Invasive mycoses in children receiving hemopoietic SCT. Bone Marrow Transplant 41(Suppl 2):S107–S111
19. Castagnola E et al (2007) A prospective study on the epidemiology of febrile episodes during chemotherapy-induced neutropenia in children with cancer or after hemopoietic stem cell transplantation. Clin Infect Dis 45(10):1296–1304
20. Castagnola E et al (2010) Incidence of bacteremias and invasive mycoses in children with acute non-lymphoblastic leukemia: results from a multi-center Italian study. Pediatr Blood Cancer 55(6):1103–1107

21. Murray PR, Rosenthal KS, Pfaller MA (2009) Medical microbiology. 6th edn. Philadelphia, PA, USA, Elsevier Mosby
22. Oz Y, Kiraz N (2011) Diagnostic methods for fungal infections in pediatric patients: microbiological, serological and molecular methods. Expert Rev Anti Infect Ther 9(3):289–298
23. Scott LJ, Simpson D (2007) Voriconazole: A review of its use in the management of invasive fungal infections. Drugs 67(2):269–298
24. Pfaller MA (1996) Nosocomial candidiasis: emerging species, reservoirs, and modes of transmission. Clin Infect Dis 22(Suppl 2):S89–S94
25. Pfaller MA et al (2000) Bloodstream infections due to Candida species: SENTRY antimicrobial surveillance program in North America and Latin America, 1997–1998. Antimicrob Agents Chemother 44(3):747–751
26. Nucci M, Anaissie E (2001) Revisiting the source of Candidemia: skin or gut? Clin Infect Dis 33(12):1959–1967
27. Strausbaugh LJ et al (1994) High frequency of yeast carriage on hands of hospital personnel. J Clin Microbiol 32(9):2299–2300
28. Pfaller MA, Diekema DJ (2007) Epidemiology of invasive candidiasis: a persistent public health problem. Clin Microbiol Rev 20(1):133–163
29. Denning DW (2003) Echinocandin antifungal drugs. Lancet 362(9390):1142–1151
30. Perea S, Patterson TF (2002) Antifungal resistance in pathogenic fungi. Clin Infect Dis 35(9):1073–1080
31. Krcmery V, Barnes AJ (2002) Non-albicans Candida spp. causing fungaemia: pathogenicity and antifungal resistance. J Hospital Infect 50(4):243–260
32. Pfaller MA et al (2006) Candida rugosa, an emerging fungal pathogen with resistance to azoles: geographic and temporal trends from the ARTEMIS DISK antifungal surveillance program. J Clin Microbiol 44(10): 3578–3582
33. Hay RJ, Jones RM (2010) New molecular tools in the diagnosis of superficial fungal infections. Clin Dermatol 28(2):190–196
34. Enoch DA, Ludlam HA, Brown NM (2006) Invasive fungal infections: a review of epidemiology and management options. J Med Microbiol 55(7):809–818
35. Rex JH et al (2000) Practice guidelines for the treatment of Candidiasis. Clin Infect Dis 30(4):662–678
36. Bicanic T, Harrison TS (2004) Cryptococcal meningitis. Br Med Bull 72(1):99–118
37. Denning DW (1998) Invasive aspergillosis. Clin Infect Dis 26(4):781–803
38. Kontoyiannis, Bodey (2002) Invasive aspergillosis in 2002: an update. Eur J Clin Microbiol Infecti Dis 21(3):161–172
39. Kontoyiannis DP, Mantadakis E, Samonis G (2003) Systemic mycoses in the immunocompromised host: an update in antifungal therapy. J Hosp Infect 53(4):243–258
40. Patterson TF et al (2000) Invasive aspergillosis: disease spectrum, treatment practices, and outcomes. Medicine 79(4):250–260
41. Haiduven D (2009) Nosocomial aspergillosis and building construction. Med Mycol 47(s1):S210–S216
42. Klont RR, Meis JFGM, Verweij PE (2001) Critical assessment of issues in the diagnosis of invasive aspergillosis. Clin Microbiol Infect 7:32–37
43. Metcalf SC, Dockrell DH (2007) Improved outcomes associated with advances in therapy for invasive fungal infections in immunocompromised hosts. J Infect 55(4):287–299
44. Nelson PE, Dignani MC, Anaissie EJ (1994) Taxonomy, biology, and clinical aspects of Fusarium species. Clin Microbiol Rev 7(4): 479–504
45. Sampathkumar P, Paya CV (2001) Fusarium infection after solid-organ transplantation. Clin Infect Dis 32(8):1237–1240
46. Thursky KA et al (2008) Recommendations for the treatment of established fungal infections. Intern Med J 38(6b):496–520
47. Greenberg RN et al (2004) Zygomycosis (mucormycosis): emerging clinical importance and new treatments. Curr Opin Infect Dis 17(6):517–525
48. Petrikkos G, Drogari-Apiranthitou M (2011) Zygomycosis in immunocompromised nonhaematological patients. Mediterr J Hematol Infect Dis 3(1):e2011012
49. De Pauw B et al (2008) Revised definitions of invasive fungal disease from the European Organization for Research and Treatment of Cancer/Invasive Fungal Infections Cooperative Group and the National Institute of Allergy and Infectious Diseases Mycoses Study Group (EORTC/MSG) Consensus Group. Clin Infect Dis 46(12):1813–1821
50. Perea S, Patterson TF (2002) Invasive Aspergillus infections in hematologic malignancy patients. Semin Respir Infect 17(2): 99–105
51. Stevens DA (2002) Diagnosis of fungal infections: current status. J Antimicrob Chemother 49(Suppl 1):11–19
52. Ellepola AN, Morrison CJ (2005) Laboratory diagnosis of invasive candidiasis. J Microbiol 43:65–84

53. Posteraro B et al (2011) Update on the laboratory diagnosis of invasive fungal infections. Mediterr J Hematol Infect Dis 3(1):e2011002
54. Sanguinetti M et al (2007) Evaluation of VITEK 2 and RapID yeast plus systems for yeast species identification: experience at a large clinical microbiology laboratory. J Clin Microbiol 45(4):1343–1346
55. Alexander BD, Pfaller MA (2006) Contemporary tools for the diagnosis and management of invasive mycoses. Clin Infect Dis 43:S15–S27
56. Marr KA, Patterson T, Denning D (2002) Aspergillosis. Pathogenesis, clinical manifestations and therapy. Infect Dis Clin North Am 16(4):875–894, vi
57. Chandrasekar P (2010) Diagnostic challenges and recent advances in the early management of invasive fungal infections. Eur J Haematol 84(4):281–290
58. Karageorgopoulos DE et al (2011) Beta-D-glucan assay for the diagnosis of invasive fungal infections: a meta-analysis. Clin Infect Dis 52(6):750–770
59. Cordonnier C et al (2009) Empirical versus preemptive antifungal therapy for high-risk, febrile, neutropenic patients: a randomized, controlled trial. Clin Infect Dis 48(8): 1042–1051
60. Almyroudis NG, Segal BH (2009) Prevention and treatment of invasive fungal diseases in neutropenic patients. Curr Opin Infect Dis 22(4):385–393
61. Dornbusch HJ, Groll A, Walsh TJ (2010) Diagnosis of invasive fungal infections in immunocompromised children. Clin Microbiol Infect 16(9):1328–1334
62. Klont RR, Mennink-Kersten MASH, Verweij PE (2004) Utility of Aspergillus antigen detection in specimens other than serum specimens. Clin Infect Dis 39(10):1467–1474
63. Maertens J et al (2001) Screening for circulating galactomannan as a noninvasive diagnostic tool for invasive aspergillosis in prolonged neutropenic patients and stem cell transplantation recipients: a prospective validation. Blood 97(6):1604–1610
64. Sanguinetti M et al (2003) Comparison of real-time PCR, conventional PCR, and galactomannan antigen detection by enzyme-linked immunosorbent assay using bronchoalveolar lavage fluid samples from hematology patients for diagnosis of invasive pulmonary aspergillosis. J Clin Microbiol 41(8):3922–3925
65. Viscoli C et al (2002) Aspergillus galactomannan antigen in the cerebrospinal fluid of bone marrow transplant recipients with probable cerebral aspergillosis. J Clin Microbiol 40(4): 1496–1499
66. Wheat LJ, Walsh TJ (2008) Diagnosis of invasive aspergillosis by galactomannan antigenemia detection using an enzyme immunoassay. Eur J Clin Microbiol Infect Dis 27(4):245–251
67. Marr KA et al (2005) Antifungal therapy decreases sensitivity of the Aspergillus galactomannan enzyme immunoassay. Clin Infect Dis 40(12):1762–1769
68. Adam O et al (2004) Treatment with piperacillin-tazobactam and false-positive Aspergillus galactomannan antigen test results for patients with hematological malignancies. Clin Infect Dis 38(6):917–920
69. Pfeiffer CD, Fine JP, Safdar N (2006) Diagnosis of invasive aspergillosis using a galactomannan assay: a meta-analysis. Clin Infect Dis 42(10):1417–1427
70. Marty FM, Koo S (2009) Role of (1->3)-beta-D-glucan in the diagnosis of invasive aspergillosis. Med Mycol 47(Suppl 1):S233–S240
71. Kohno S et al (1993) An evaluation of serodiagnostic tests in patients with candidemia: beta-glucan, mannan, candida antigen by Cand-Tec and D-arabinitol. Microbiol Immunol 37(3):207–212
72. Ostrosky-Zeichner L et al (2005) Multicenter clinical evaluation of the (1->3) beta-D-glucan assay as an aid to diagnosis of fungal infections in humans. Clin Infect Dis 41(5):654–659
73. Odabasi Z et al (2004) Beta-D-glucan as a diagnostic adjunct for invasive fungal infections: validation, cutoff development, and performance in patients with acute myelogenous leukemia and myelodysplastic syndrome. Clin Infect Dis 39(2):199–205
74. Morrissey CO et al (2008) Diagnostic and therapeutic approach to persistent or recurrent fevers of unknown origin in adult stem cell transplantation and haematological malignancy. Intern Med J 38(6b):477–495
75. Weiner MH, Coats-Stephen M (1979) Immunodiagnosis of systemic candidiasis: mannan antigenemia detected by radioimmunoassay in experimental and human infections. J Infect Dis 140(6):989–993
76. Yera H et al (2001) Contribution of serological tests and blood culture to the early diagnosis of systemic candidiasis. Eur J Clin Microbiol Infect Dis 20(12):864–870
77. Pasqualotto AC, Denning DW (2005) Diagnosis of invasive fungal infections – current limitations of classical methods and new diagnostic methods. Europ Oncol Rev REV:1–11
78. Hauggaard A, Ellis M, Ekelund L (2002) Early chest radiography and CT in the diagnosis, management and outcome of invasive

pulmonary aspergillosis. Acta Radiol 43(3):292–298
79. Primack SL et al (1994) Pulmonary nodules and the CT halo sign. Radiology 190(2): 513–515
80. Pinto PS (2004) The CT halo sign. Radiology 230(1):109–110
81. Caillot D et al (2001) Increasing volume and changing characteristics of invasive pulmonary aspergillosis on sequential thoracic computed tomography scans in patients with neutropenia. J Clin Oncol 19(1):253–259
82. Bader O et al (2011) Improved clinical laboratory identification of human pathogenic yeasts by matrix-assisted laser desorption ionization time-of-flight mass spectrometry. Clin Microbiol Infect 17:1359–1365
83. Alanio A et al (2011) Matrix-assisted laser desorption ionization time-of-flight mass spectrometry for fast and accurate identification of clinically relevant Aspergillus species. Clin Microbiol Infect 17(5):750–755
84. White PL, Perry MD, Barnes RA (2009) An update on the molecular diagnosis of invasive fungal disease. FEMS Microbiol Lett 296(1):1–10
85. White PL et al (2006) The evolution and evaluation of a whole blood polymerase chain reaction assay for the detection of invasive aspergillosis in hematology patients in a routine clinical setting. Clin Infect Dis 42(4): 479–486
86. Dunyach C et al (2008) Detection and identification of Candida spp. in human serum by LightCycler real-time polymerase chain reaction. Diagn Microbiol Infect Dis 60(3):263–271
87. Costa C et al (2002) Real-time PCR coupled with automated DNA extraction and detection of galactomannan antigen in serum by enzyme-linked immunosorbent assay for diagnosis of invasive aspergillosis. J Clin Microbiol 40(6):2224–2227
88. Landlinger C, Preuner S, Bašková L, van Grotel M, Hartwig NG, Dworzak M, Mann G, Attarbaschi A, Kager L, Peters C, Matthes-Martin S, Lawitschka A, van den Heuvel-Eibrink MM, Lion T (2010) Diagnosis of invasive fungal infections by a real-time panfungal PCR assay in immunocompromised pediatric patients. Leukemia 24:2032–2038
89. Khot PD, Fredricks DN (2009) PCR-based diagnosis of human fungal infections. Expert Rev Anti Infect Ther 7(10):1201–1221
90. Barnes RA (2008) Early diagnosis of fungal infection in immunocompromised patients. J Antimicrob Chemother 61(Suppl 1):i3–i6
91. Donnelly JP (2006) Polymerase chain reaction for diagnosing invasive aspergillosis: getting closer but still a ways to go. Clin Infect Dis 42(4):487–489
92. Buchheidt D et al (2004) Prospective clinical evaluation of a LightCycler-mediated polymerase chain reaction assay, a nested-PCR assay and a galactomannan enzyme-linked immunosorbent assay for detection of invasive aspergillosis in neutropenic cancer patients and haematological stem cell transplant recipients. Br J Haematol 125(2):196–202
93. Wheat LJ (2006) Antigen detection, serology, and molecular diagnosis of invasive mycoses in the immunocompromised host. Transpl Infect Dis 8(3):128–139
94. Mengoli C et al (2009) Use of PCR for diagnosis of invasive aspergillosis: systematic review and meta-analysis. Lancet Infect Dis 9(2):89–96
95. Archibald L et al (1997) Antimicrobial resistance in isolates from inpatients and outpatients in the United States: increasing importance of the intensive care unit. Clin Infect Dis 24(2):211–215
96. Ghimire A, Dela Cruz CS (2011) Fungal infections in the intensive care unit. The American College of Chest Physicians. PCCSU. Vol 25
97. Ruping MJGT, Vehreschild JJ, Cornely OA (2008) Patients at high risk of invasive fungal infections when and how to treat. Drugs 68(14):1941–1962
98. Viscoli C (2009) Antifungal prophylaxis ans preemptive therapy. Drugs 69(Suppl 1):75–78
99. Playford EG, Lipman J, Sorrell TC (2010) Management of invasive candidiasis in the intensive careunit. Drugs 70:823–839
100. Rubio-Terres C, Grau S (2010) Pharmacoeconomics of voriconazole. Expert Opin Pharmacother 11(6):877–887
101. Ashley ESD et al (2006) Pharmacology of systemic antifungal agents. Clin Infect Dis 43: S28–S39
102. Loeffler J, Stevens DA (2003) Antifungal drug resistance. Clin Infect Dis 36(Suppl 1): S31–S41
103. Mohr J et al (2008) Current options in antifungal pharmacotherapy. Pharmacotherapy 28(5):614–645
104. Gallis HA, Drew RH, Pickard WW (1990) Amphotericin B: 30 years of clinical experience. Rev Infect Dis 12(2):308–329
105. Boucher HW et al (2004) Newer systemic antifungal agents: pharmacokinetics, safety and efficacy. Drugs 64(18):1997–2020

106. Ng AW, Wasan KM, Lopez-Berestein G (2003) Development of liposomal polyene antibiotics: an historical perspective. J Pharm Pharm Sci 6(1):67–83
107. Kontoyiannis DP et al (2003) Sequential exposure of Aspergillus fumigatus to itraconazole and caspofungin: evidence of enhanced in vitro activity. Diagn Microbiol Infect Dis 47(2):415–419
108. Kauffman CA (2006) Fungal infections. Proc Am Thorac Soc 3(1):35–40
109. Fluckiger U et al (2006) Treatment options of invasive fungal infections in adults. Swiss Med Wkly 136(29–30):447–463
110. McCoy D, Depestel DD, Carver PL (2009) Primary antifungal prophylaxis in adult hematopoietic stem cell transplant recipients: current therapeutic concepts. Pharmacotherapy 29:1306–1325
111. Lai CC et al (2008) Current challenges in the management of invasive fungal infections. J Infect Chemother 14(2):77–85
112. Letscher-Bru V, Herbrecht R (2003) Caspofungin: the first representative of a new antifungal class. J Antimicrob Chemother 51(3):513–521
113. Auberger J, Lass-Flörl C (2011) Current evidence for the treatment of invasive fungal infections in immunocompromised patients. Clin Infect 1(3):447–457
114. Richardson MD, Jones BL (2007) Therapeutic guidelines in systemic fungal infections. 3rd edn. Current Medical Literature.
115. Safdar A (2007) Antifungal immunity and adjuvant cytokine immune enhancement in cancer patients with invasive fungal infections. Clin Microbiol Infect 13(1):1–4
116. İkincioğulları A et al (2005) Granulocyte transfusions in children with chronic granulomatous disease and invasive aspergillosis. Ther Apher Dial 9(2):137–141
117. Ghannoum MA, Rice LB (1999) Antifungal agents: mode of action, mechanisms of resistance, and correlation of these mechanisms with bacterial resistance. Clin Microbiol Rev 12(4):501–517
118. Kontoyiannis DP, Lewis RE (2002) Antifungal drug resistance of pathogenic fungi. Lancet 359(9312):1135–1144
119. Sheehan DJ, Hitchcock CA, Sibley CM (1999) Current and emerging azole antifungal agents. Clin Microbiol Rev 12(1):40–79
120. Arikan S (2007) Current status of antifungal susceptibility testing methods. Med Mycol 45(7):569–587

Chapter 2

Fungal Diagnostics: Review of Commercially Available Methods

Javier Yugueros Marcos and David H. Pincus

Abstract

Fungi and yeasts are critical causes of acute infection. As such, the detection and identification of these organisms are crucial in the diagnosis of affected patient populations. There is a vast array of commercial tests currently available for diagnostic purposes. These vary from traditional culture and biochemical methods to advanced multiparameter molecular tests. Recent technological advances have driven the development of rapid tests which are complementing and in some cases replacing the more traditional methods of detection. Irrespective of the method used the ultimate goal is timely detection of the infectious agent allowing appropriate treatment and improved outcome for the patient.

Key words: Commercial test, Fungal diagnostics, Direct examination, Chromogenic media, Yeast identification, Antigen-based test, Antibody-based test, Nucleic acid probes, Molecular biology

1. Introduction

Fungi have progressively emerged as an important cause of life-threatening infections. Patient populations at greatest risk are those critically ill and/or immunocompromised, including surgical intensive care unit (ICU), advanced human immunodeficiency virus (HIV) infected, inherited immunodeficiency-prolonged neutropenia, and solid organ and hematopoietic stem cell transplant (HSCT) recipients (1, 2).

Significant progress has been made over the last 10 years in the field of fungal detection and identification. Technological advances and introduction of new technologies have led to availability of a wide variety of commercial tests. Their use in the clinical setting has been successful, some moving into clinical

Louise O'Connor and Barry Glynn (eds.), *Fungal Diagnostics: Methods and Protocols*, Methods in Molecular Biology, vol. 968, DOI 10.1007/978-1-62703-257-5_2, © Springer Science+Business Media New York 2013

Table 1
Common histological stains used for diagnosis of fungal infections

Stain[a]	Fungal elements	Background
GMS	Brown to black	Green to blue
PAS	Pink to red	Pale pink to pale purple
GF	Pink to red to purple	Yellow
H&E	Pink to red to purple	Pink

[a] *GMS* Gomori methenamine silver, *PAS* periodic acid-Schiff, *GF* Gridley fungus, *H&E* hematoxylin and eosin

guidelines for fungal infection management. This chapter reviews commercially available methods for detection and identification of the fungi most commonly associated with infection. The aim is to provide the reader with an overview of the extensive range of methods available focusing on tests, kits, assays, or systems for use in Europe, USA, or Japan.

2. Direct Examination of Clinical Samples

Direct examination of clinical samples for fungi is made possible by several stains enabling a more obvious appearance of fungal elements. Classical stains used in histopathology include Gomori methenamine silver, periodic acid-Schiff, Gridley fungus, and hematoxylin and eosin stains (3) (Table 1). Alternatively, calcofluor white (CW) can be used with a fluorescent microscope to observe fungal elements in clinical samples. CW binds β-glycosidic linkages of polysaccharides in the fungal cell wall but also binds nonspecifically to keratin and human connective tissue elements (4).

3. Culture-Based Methods for Fungal Detection

3.1. Nonspecific Detection of Fungi

Culture media used to cultivate fungi include various non-selective agars such as Sabouraud dextrose agar (SDA), potato dextrose agar, and brain heart infusion (BHI) agar. In order to inhibit growth of bacteria or saprophytic fungi, selective agents such as chloramphenicol, gentamicin, and cycloheximide are added. Broths (e.g., Sabouraud dextrose broth, BHI broth, etc.) are used occasionally to grow or enrich for fungi. Most fungi are non-fastidious and grow on most media (including trypticase soy blood agar) that supply sources of nitrogen (e.g., peptone) and carbon (e.g., glucose). The disadvantage of these media is that many fungi look alike and additional methods and more incubation are needed for identification. That delay in diagnosis was the impetus for development of more specific methods such as the chromogenic media which will be described later in this review.

Automated blood culture systems are also available and include the BACTEC™ (BD™ Diagnostic Systems, Sparks, MD, USA), BacT/Alert® 3D (bioMérieux®, Marcy l'Etoile, France), and VersaTREK® (TREK Diagnostic Systems, Cleveland, OH, USA). Aside from using different broth components and/or volumes, these systems employ different detection methods to monitor growth and metabolism, i.e., CO_2 generation and subsequent pH decreases detected by fluorimetric BACTEC™ (5) or colorimetric BacT/Alert® 3D (6) sensors or headspace gas changes detected by VersaTREK®'s pressure sensors (7).

3.2. Specific Detection of Fungi

The germ tube test described in 1960 (8) was the main method for rapid identification of *C. albicans* until the advent of enzymatic tests and chromogenic media. Some labs still use the germ tube test for rapid inexpensive detection of *C. albicans* but as it relies on microscopic observation and technical expertise, it has been largely displaced by enzymatic tests. Table 2 shows commercial tests that use a combination of substrates to detect presence of *N*-acetyl-β-D-galactosaminidase and L-proline arylamidase. While

Table 2
Rapid (30–60 min) identification of *Candida albicans* using NGL/PRO[a]

Test	Vendor	No. isolates tested	% SN	% SP	Reference
BactiCard® Candida	Remel, Lenexa, KS, USA	583	99.3	99.6	(9)
		133	100.0	100.0	(10)
		536	97.0[b]	97.1[b]	(11)
		196	97.8	92.5	(12)
Candida albicans Test Kit (MUREX *C. albicans* 50)	Remel	583	98.7	99.6	(9)
		502	94.6	97.8	(13)
		133	100.0	100.0	(10)
O.B.I.S. albicans	Oxoid, Basingstoke, UK	219	100.0	100.0	(14)
AlbiQuick™ Test Kit	Hardy Diagnostics, Santa Maria, CA, USA	NR	NR	NR	NR

[a] *NGL* β-galactosaminidase, *PRO* proline arylamidase, *SN* sensitivity, *SP* specificity, *NR* not reported, *TP* true positive
[b] Recalculated to include *Candida dubliniensis* as TP

both enzymes are present in *C. albicans*, they are also found in some less common species, e.g., *C. dubliniensis*, *C. rugosa*, and *Trichosporon* spp.

Staib (15) first described birdseed (*Guizotia abyssinica*) agar as a useful screening tool for detection and rapid identification of *C. neoformans*. Most strains of *C. neoformans* are detected within 72 h incubation on this medium where phenoloxidase activity results in dark brown colored colonies. A comparison of birdseed agar to conventional media (SDA and Mycosel™ agar; BD™ Diagnostic Systems) with 35 clinical samples from AIDS patients showed 100% sensitivity and specificity with plates incubated at 30°C. Authors suggested that this or a higher temperature (37°C) could be more selective and avoid rare false-positives reported for non-*C. neoformans* cryptococci. Although no problems have been reported with bacterial overgrowth, antibiotic supplementation has been suggested to aid screening samples with heavy background bacterial flora (16). Another commercially available medium used for identification of *C. neoformans* and *C. albicans*, within 24 h from previously isolated colonies, is TOC (tween 80-oxgall-caffeic acid) agar (17). However, its use as a primary isolation medium requires extended incubation of 3–5 days. Other investigators (18, 19) recommended urease for rapid recognition of *C. neoformans*, but this lacks specificity and needs to be followed by a more reliable method. Rapid urease can be useful for screening respiratory isolates when the target species are limited to *C. neoformans* and *C. gattii*.

The rapid trehalose assimilation test and its modifications can be used for rapid identification of *C. glabrata* (Table 3). Rapid identification of *C. glabrata* is critical due to its resistance or reduced susceptibility to triazole antifungals, e.g., fluconazole (28).

3.3. Chromogenic Media

Chromogenic media allow direct identification of the most common clinical *Candida* species. They also allow for recognition of mixed yeast cultures that may be difficult or impossible to see on conventional mycological media, e.g., SDA. They incorporate substrates, e.g., halogenated indoxylic derivatives, used to detect specific enzymes. Substrate cleavage results in formation of brightly colored indigo dyes that are retained within yeast cells resulting in specific color development in colonies of target species. These media also contain one or more antibiotics used to prevent growth of contaminating bacterial flora. *Candida albicans* is identified by β-hexosaminidase activity, which is also found in the less common and closely related species *C. dubliniensis*. Some of these media, e.g., CHROMagar™ Candida (CHROMagar™, Paris, France), allow for presumptive identification of other yeasts, e.g., *C. tropicalis*, through detection of additional enzymes, e.g., phosphatase (29). In this case, *C. tropicalis* forms blue colonies by hydrolysis of both chromogenic substrates. Presumptive identification may be

Table 3
Rapid identification of *Candida glabrata*[a]

Test	Vendor	No. isolates tested	% SN	% SP	Reference
Glabrata RTT	Fumouze Diagnostics, Levallois-Perret Cedex France	330	≥94.0	≥97.3	(20)[b]
		1,174	95.8	98.9	(21)
		168	98.4	100.0	(22)
		332	≥91.1	≥83.9	(23)[c]
Rapid Trehalose Assimilation Broth	Remel	320	91.5	96.3	(24)
Rapid trehalose + sucrose assimilation with Rosco disks (Diatabs™)	Rosco Diagnostica A/S, Taastrup, Denmark	440	100.0	100.0	(25)
Rapid trehalase (solution from Diatabs™ w/Clinistix®)	Rosco Diagnostica A/S; Bayer, Tarrytown, NY, USA	482	≥98.2	≥95.3	(26)
Rapid trehalase/maltase	Rosco/Bayer	255	96.9	98.4	(26)
RAT (Mayo Clinic method)	Scientific Device Laboratory, Des Plaines, IL, USA	91	≥81.3	100.0	(27)[d]
GlabrataQuick™ Kit	Hardy Diagnostics	NR	NR	NR	NR

[a]*SN* sensitivity, *SP* specificity, *NR* not reported
[b]Four isolation media tested; % SN range was 94.0–98.0 and % SP range was 97.3–98.6
[c]Five isolation media tested; % SN range was 91.1–96.7 and % SP range was 83.9–99.6
[d]Two isolation media tested; % SN range was 81.3–100.0 and % SP was 100.0

possible for other species, e.g., *C. krusei* or *C. glabrata*, but performance is related to expertise in recognition of morphologic or discrete color differences from other similar species (30).

Chromogenic media available for the detection of Candida species include Brilliance™ Candida (Oxoid, Basingstoke, UK) (31), Candi*Select*™ 4 (Bio-Rad, Marnes-la-Coquette, France) (32), CHROMagar™ Candida (33, 34), and chromID™ Candida (bioMérieux®, Marcy l'Etoile, France) (35).

4. Post-culture Identification Methods

All of the non-molecular commercial identification systems described in Subheadings 4.1–4.3 require inoculation with a pure culture isolate while the molecular methods in Subheading 4.4 allow processing from positive blood culture bottles. The different methods relying on pure culture isolates have varying culture requirements, i.e., isolation medium, culture age, and inoculum suspension density.

4.1. Manual Identification Systems

The early manual identification systems on the market were limited in the scope of species that could be identified, which was a function of limited number of substrates and immature databases. As commercial methods evolved, so did the numbers/types of substrates and their databases. Table 4 shows a summary of the available manual yeast identification methods. For the purpose of consistency, some species and/or biotype claims listed in the various product package inserts were abbreviated in order to keep in line with current taxonomy, i.e., *C. albicans/C. stellatoidea*, *C. albicans* 1/*C. albicans* 2, *C. paratropicalis/C. tropicalis*, etc. that are conspecific were counted as single claims.

4.2. Automated Identification Systems

With the introduction of automation, workflow became easier, and reaction and final result interpretations became objective and more reproducible. These advances allowed accurate differentiation of a much wider range of clinically relevant taxa. Table 5 summarizes the systems available commercially.

4.3. Matrix-Assisted Laser Desorption Ionization: Time of Flight Mass Spectrometry

Matrix-assisted laser desorption ionization–time of flight mass spectrometry (MALDI-TOF MS) is one of the newest methods used for fungal identification. This is a very accurate and rapid technique that requires minimal expertise or hands-on time. The method is based primarily on ribosomal proteins. Whole cell preparations or extracts are embedded in a chemical matrix and ionized by a laser. The resultant molecules migrate through a charged field in a vacuum tube toward a detector. The differences in time of flight for these ions translate to different mass peaks in the range of 2–20 kDa. These spectra are then compared to a database of known species to arrive at an identification within minutes of the ionization step. Results have shown that this method is less dependent on culture conditions than many other commercial phenotypic methods and therefore allows for fairly broad ranges of growth conditions, i.e., isolation media, incubation time, and temperature.

There are currently two commercial systems available the BioTyper MALDI-TOF (microflex™ LT instrument; Bruker Daltonics, Bremen, Germany) and Vitek® MS™ (bioMérieux®).

BioTyper MALDI-TOF: As of the end of 2011, this system had a European Conformity (CE; Conformité Européene) mark and could be used for in vitro diagnostic (IVD) reporting in European Union countries. In other countries, e.g., USA, the format was research use only (RUO) prior to other regulatory, e.g., Food and Drug Administration (FDA), approvals.

Vitek® MS™: This MALDI-TOF MS system uses similar principles to the Bruker system and became available in early 2011 with the CE IVD mark. Prior to development of commercial software and other regulatory approvals, this system also used an RUO

Table 4
Manual comprehensive yeast identification methods

Product (vendor)	No. strains	% Correct	% NoID	% MisID	Author
api® Candida (bioMérieux®)	609	97.4	0.7	2.0	Fricker-Hidalgo et al. (36)
	198	91.4	5.1	3.5	Bernal et al. (37)
	156	82.7	4.5	12.8	Buchaille et al. (38)
	159	91.8	0.6	7.5	Campbell et al. (39)
	202	97.0	2.5	0.5	Paugam et al. (40)
api® 20C AUX (bioMérieux®)	505	96.2	0.0	3.8	Buesching et al. (41)
	1,093	97.3	2.7	0.0	Land et al. (42)
	126	77.0	19.8	3.2	Bergan et al. (43)
	178	89.3	5.6	5.1	Schuffenecker et al. (44)
	206	94.7	0.0	5.3	Willemsen et al. (45)
	243	93.0	6.2	0.8	Ramani et al. (46)
	171	96.5	1.2	2.3	Sand and Rennie (47)
	201	100.0	0.0	0.0	Wadlin et al. (48)
	116	99.1	0.0	0.9	Gündes et al. (49)
Auxacolor™ (Bio-Rad)	178	88.8	5.6	5.6	Schuffenecker et al. (44)
	97	79.4	15.5	5.2	Milan et al. (50)
	206	62.1	NR	NR	Willemsen et al. (45)
	156	85.9	10.3	3.8	Buchaille et al. (38)
	105	91.4	4.8	3.8	Sheppard et al. (51)
	159	91.2	8.2	0.6	Campbell et al. (39)
	202	94.1	4.9	1.0	Paugam et al. (40)
	100	94.0	6.0	0.0	Romney et al. (52)
Candifast® (ELITech Group)	178	51.1	25.3	23.6	Schuffenecker et al. (44)
	116	88.8	5.2	6.0	Gündes et al. (49)
Fungichrom® (ELITech Group)	156	84.6	10.9	4.5	Buchaille et al. (38)
	202	92.1	5.4	2.5	Paugam et al. (40)
	116	95.7	2.6	1.7	Gündes et al. (49)
Fungifast® (ELITech Group)	156	77.6	19.9	2.6	Buchaille et al. (38)
	202	88.1	10.9	1.0	Paugam et al. (40)
ID 32 C (bioMérieux®)	69	76.8	0.0	23.2	Latouche et al. (53)
	156	98.1	1.9	0.0	Buchaille et al. (38)
	243	88.9	10.3	0.8	Ramani et al. (46)
	250	93.6	0.8	5.6	Marklein et al. (54)
RapID™ Yeast Plus (Remel)	300	99.3	0.0	0.7	Kitch et al. (55)
	156	78.2	5.8	16.0	Buchaille et al. (38)
	447	90.6	2.5	6.9	Espinel-Ingroff et al. (56)
	133	94.0	0.0	6.0	Heelan et al. (57)
	201	99.0	0.0	1.0	Wadlin et al. (48)
	750	95.5	2.4	2.1	Sanguinetti et al. (58)
Uni-Yeast-Tek® (Remel)	623	99.8	0.0	0.2	Bowman and Ahearn (59)
	229	99.1	0.0	0.9	Bowman and Ahearn (60)
	206	94.2	NR	NR	Cooper et al. (61)
	126	88.7	0.0	11.3	Bergan et al. (43)
	489	37.2	49.7	13.1	Salkin et al. (62)

Table 5
Automated comprehensive yeast identification methods

Product (vendor)	No. strains	% Correct	% NoID	% MisID	Reference
Biolog YT Microplate™ (Biolog)	129	48.8	12.4	38.8	(63)
	171	48.5	8.8	42.7	(47)
MicroScan® Rapid YS (Siemens)	437	85.4	NR	8.2	(64)
	357	96.6	0.6	2.8	(65)
	150	85.3	NR	NR	(66)
Sherlock® MIS (MIDI, Inc.)	550	68.0	16.2	15.8	(67)
	477	70.2	6.1	23.7	(68, 69)
Vitek® YBC (bioMérieux®)	253	95.7	0.0	4.3	(70)
	352	84.9	0.0	15.1	(80)
	1,106	93.4	NR	NR	(71)
	221	83.3	2.7	14.0	(81)
	398	97.2	1.3	1.5	(72)
	222	80.6	14.4	5.0	(74)
	150	95.0	NR	NR	(66)
	409	89.7	3.2	7.1	(73)
	55	87.3	0.0	12.7	(53)
	171	96.5	0.6	2.9	(47)
	201	91.0	4.0	5.0	(48)
Vitek® 2 YST (bioMérieux®)	97	99.0	0.0	1.0	(75)
	172	93.6	3.5	2.9	(76)
	623	98.6	0.5	0.9	(77)
	750	98.4	0.5	1.1	(58)
	136	94.1	1.2	4.7	(78)
	68	97.1	0.0	2.9	(79)

format. Pre-2011 publications address studies of this RUO format (SARAMIS™ software) and although most evaluated bacterial identification, fungi were also tested.

4.4. Tests Utilizing Nucleic Acids for Fungal Detection

FISH has routinely been employed as a method of detection of fungi for some time. One commercial system which utilizes the method is the Yeast Traffic Light® PNA-FISH® (AdvanDx, Woburn, MA, USA). This is a peptide nucleic acid fluorescent in situ hybridization (PNA FISH) method employing fluorescein-labeled probes that hybridize with 26S ribosomal RNA (rRNA) of target species, e.g., *C. albicans*, that can be identified directly using smears of positive blood cultures. Other nucleic acid based methods using cultured materials as sample include Prove-it™ Fungi and BlackLight® Fungal ID kit.

Prove-it™ Fungi (Mobidiag, Helsinki, Finland) is a rapid DNA-based test intended for identification of 13 yeast species:

C. albicans, *C. tropicalis*, *C. glabrata*, *C. parapsilosis*, *C. krusei*, *C. lusitaniae*, and *C. guilliermondii* as well as six more species (*C. dubliniensis*, *C. pelliculosa*, *C. kefyr*, *C. norvegensis*, *C. haemulonii*, and *Saccharomyces cerevisiae*) enclosed in the so-called pan-yeast panel. The method, based on extraction of DNA from either positive blood culture bottles or pure culture samples, uses polymerase chain reaction (PCR) and a small microarray placed at the bottom of the tube, to detect fluorescence signals emitted when specific yeast sequences hybridize. The total turnaround time is approximately 3 h once DNA is extracted and purified.

The BlackLight® Fungal ID kit (2B BlackBio S.L., Madrid, Spain) applies the principle of pyrosequencing, which relies on detection of pyrophosphate release at each nucleotide incorporation during sequencing reaction. The test starts from a 15 μl dry blood spot (1 mm diameter), which is used directly for PCR without pretreatment. The subsequent sequencing of a small region of the 18S rRNA gene allows the test to identify the most relevant *Candida* species as well as *Aspergillus fumigatus* and *A. niger*.

The assay is available in Europe for IVD purposes and at the moment this review was written, no clinical evaluation was yet published. The proof-of-principle description and data on analytical performance of the assay were discussed in a recent paper, where they showed 100% agreement with standard microbiological methods, when testing a collection of clinical blood culture isolates (82).

5. Antigen and Antibody-Based Methods for Fungal Detection

Detection of antigenic compounds from the fungal cell wall (e.g., chitin, β-glucans, and mannoproteins) or the respective antibodies generated by the host immune system is commonly used, in conjunction with other diagnostic procedures (e.g., microbiological cultures, histological examination of biopsy samples, and radiographic evidence), as an aid in the diagnosis of fungal infections.

5.1. Antigen- and Antibody-Based Tests for Candida

Mannan, a polysaccharide non-covalently bound to the yeast cell wall, can act as an antigen, generating an immune response that leads to production of host antibodies (83). Detection of mannan or respective antibodies led to commercialization of one of the most widely used tests for the laboratory detection of *Candida*: Platelia™ Candida Ag and Platelia™ Candida Ab (Bio-Rad). These are sandwich enzyme-linked immunosorbent assay (ELISA) tests for quantitative detection of *Candida* mannan or anti-mannan antibodies, in serum of subjects at-risk or suspected of invasive candidiasis. Both assays have been available in Europe for about 10 years, but neither has received FDA approval for USA IVD use.

Overall sensitivity of mannan detection seems to be in the range of 60% and specificity around 90%, according to analysis of 12 published studies on adult populations (84). Similar to better figures were also found in pre-term infants in a neonatal ICU, suggesting that regular serology surveillance of circulating mannan would be a good complement to blood cultures for early detection of invasive candidiasis in this particular population (85).

Due to rapid elimination of mannan during infection, a negative result from the Platelia™ Candida Ag test cannot rule out diagnosis of invasive candidiasis. The manufacturer recommends concomitant screening of anti-mannan antibodies, by means of the Platelia™ Candida Ab assay, a two-step indirect immunoenzymatic microplate assay that allows quantitative detection of antibodies in human serum (86). This guidance is supported by several studies and literature analysis, reporting sensitivity improvements up to 85–90% (84, 87–89), and was recently endorsed by the experts report from the 3rd European Conference on Infections in Leukemia (90).

Other tests and kits based on the use of mannan and other specific antigens for *Candida* detection have been developed and commercialized by different companies. Serion Immunodiagnostica GmbH (Würzburg, Germany) commercializes the Serion ELISA antigen Candida and the Serion ELISA classic *Candida albicans* IgG/IgM/IgA, two immunoassays for either quantitative or qualitative use. The first is based on a mixture of cytoplasmic antigen structures such as enolase, and a certain amount of other cell-wall components, allowing detection of *Candida* infection in serum or plasma. The second is aimed to detect human antibodies directed against *C. albicans.* Very few studies describing and comparing performance of these assays have been published to date (91–93). Wulf and coworkers showed that the antibody-based test could pinpoint evidence for invasive candidiasis significantly earlier (approximately 22 days) (93).

Indirect tests for *Candida* detection based on the principle of latex agglutination (LA) are also commercially available. The Pastorex™ Candida (Bio-Rad) uses latex particles coated with specific antibodies that react with the polysaccharide antigen from *Candida* through agglutination, making it visible to the naked eye. Even though clinical specificity is pretty high, the clinical sensitivity remains quite low, around 25% (94, 95).

The Cand-Tec™ (Ramco Laboratories Inc. Stafford, TX, USA) is another agglutination assay for *Candida* detection in human samples. One retrospective study compared its sensitivity and specificity with the Pastorex™ Candida assay, in a very limited cohort of patients. While sensitivity values for Cand-Tec™ were much better (68.8% vs. 16.7%), specificity was significantly lower (57.1% vs. 100%) (95). The use of this assay in combination with indirect hemagglutination seems to improve clinical performance in ICU patients (sensitivity 100% and specificity 83.3%) (96).

Based on the principle of indirect immunofluorescence and targeting a mannoprotein located on the germ tube cell wall surface, Vircell Microbiologists (Granada, Spain) commercializes an assay for detection of IgG antibodies against *C. albicans* germ tube antigens (CAGTA). This assay showed sensitivity of 83–85% and specificity of 86–95% through different studies (97, 98). Titers can be lower in case of invasive infections caused by *Candida* species other than *C. albicans*, and also in immunocompromised patients, even though overall performance does not seem affected (98, 99). Interestingly, a recent study suggested its potential value as a prognostic marker, and although further investigations are needed, their findings did show significantly lower mortality rates in patients with increased CAGTA titers, concomitantly treated with antifungal agents (100).

5.2. Antigen- and Antibody-Based Tests for Cryptococcus

Cryptococcal antigen detection has shown very high sensitivity and specificity, making it very helpful and reliable for diagnosis of cryptococcal pulmonary infection and meningitis.

Several commercial methods are available. The specificity of most cryptococcal antigen tests can be obscured by presence of rheumatoid interference factors in specimens, as they can lead to false-positive reactions. This is especially true for serum samples but much less for cerebrospinal fluid (CSF) specimens, which rarely contain antiglobulins responsible for this cross-reaction. For this reason, most manufacturers (e.g., IMMY, Meridian Biosciences Inc., and Bio-Rad) propose and recommend use of pronase. Pretreatment of biological specimens with pronase was shown to significantly reduce nonspecific interference of noisy substances and consequently, reduce false-positive results (101, 102). Additional factors that could lead to false-positive results are infection by *Trichosporon* species, use of starch, some disinfectants, and soap (103), and use of BBL™ Port-A-Cul™ anaerobic vials for specimen transport (104). Furthermore, it is recommended to process CSF specimens prior to culture as immersion of a platinum wire/loop containing trace amounts of surface condensation from agar has been described as a source of interference (105). Sample storage is also important, as increased pH and high temperature induced degradation of cryptococcal antigen and therefore loss of reactivity for serological tests (106). Commercially available tests for *Cryptococcus* species are summarized in Table 6.

5.3. Antigen- and Antibody-Based Tests for Aspergillus

Detection of circulating galactomannan (GM) is generally utilized for diagnosis and management of invasive aspergillosis. GM is a heteropolysaccharide component of the cell walls of *Aspergillus*

Table 6
Commercially available antigen or antibody-based tests for *Cryptococcus* detection

Format	Assay name	Manufacturer	Reference
Latex agglutination	Cryptococcal Antigen Latex Agglutination System (CALAS®)	Meridian Bioscience, Inc. Cincinnati, OH, USA	(105–107)
			Not reported
	Latex-Cryptococcus Antigen Test	IMMY	(99, 108)
	Pastorex™ Crypto Plus	Bio-Rad	(106, 107)
	Cryptococcus Antigen Test	Remel	(107)
	Crypto-LA®	Wampole Laboratories Inc., Cranbury, NJ, USA	(108, 109)
	Eiken Latex test	Eiken, Tokyo, Japan	
Enzymatic Immuno-Assays	Premier™ Cryptococcal Antigen	Meridian Bioscience Inc.	(105, 107)

and *Penicillium* species and was the first circulating antigen found in animal models with invasive aspergillosis. The Platelia™ Aspergillus EIA (enzyme immunoassay; Bio-Rad) is a one-stage immunoenzymatic sandwich microplate assay which uses rat monoclonal antibody EBA-2, for detection of *Aspergillus* GM antigen in serum. CE-IVD marked by the end of the 1990s, it was also FDA approved in 2003. Three recent publications review its use, performance variability, and assay limitations, providing guidance for better results interpretation (110–112). A wide variety of factors affect performance of Platelia™ Aspergillus EIA including host conditions, concomitant treatments, presence of other fungal species, handling and processing of biological specimens, etc. Even though other authors provided more exhaustive lists and further description of false-positive and false-negative results (111–113), Table 7 highlights some of the most common sources of interference that can confer misleading results. Measurement of GM is currently recommended by the international guidelines for management of patients with invasive aspergillosis, where it is clearly highlighted that clinical judgment must prevail when interpreting test results (2).

Even though the manufacturer does still not claim use of the Platelia™ Aspergillus EIA for samples other than serum, several different reports since 2004 pointed out the interest of GM measurement in other specimens, e.g., urine, BAL, or CSF, because of higher sensitivity and potential early detection over the course of infection in immunocompromised patients (114–118). However, its use in BAL samples from immunocompetent patients seems to have no added value (119). The utility of the Platelia™ Aspergillus EIA in pediatric patients is still unclear and contradictory results

Table 7
Most common variables associated with false-positive and false-negative results for the Platelia™ Aspergillus EIA test (Bio-Rad)

Factor and/or situation that can lead to false-positive results	
Host related	Renal failure Mucositis Food intake of galactofuranose[a] Gut colonization and potential translocation of *Bifidobacterium* Gastrointestinal microflora of neonates
Iatrogenic	Blood derivatives Intravenous solutions containing gluconate Treatment with antibiotics derived from the fermentation of *Penicillium* species (e.g., piperacilin-tazobactam, amoxicilin-clavulanic acid) Use of cyclophosphamide in cancer patients
Sample collection and/or processing	Use of materials such as cotton swabs and cardboard Inappropriate cut-off value (too low)
Environmental	Presence of other non-*Aspergillus* fungi such as *Penicillium*, *Alternaria*, *Paecilomyces*, *Geotrichum*, *Histoplasma*, and even *C. neoformans*[b]
Factor and/or situation that can lead to false-negative results	
Host conditions	Chronic granulomatose disease
Iatrogenic	Treatment with antifungals
Sample collection and/or processing	Long-term storage of samples Inappropriate cut-off value (too high)

[a]Stabilizing agent commonly used in some cereal-derived products and some cream desserts
[b]Epitope contained within its galactoxylomannan has been shown to cross-react (113)

have been reported among different studies. Although still not recommended by the manufacturer further validation studies might help to clarify its positioning within this critical population (120). In all cases, the significance of positive results in patients without confirmed aspergillosis always requires further evaluation.

Another assay format to detect GM in serum is proposed with Pastorex™ Aspergillus (Bio-Rad), a qualitative test, which uses a simple agglutination technique. The lowest concentration of GM detectable is 15 ng/ml, around ten times more than the 1 ng/ml claimed by the Platelia™ Aspergillus EIA. Definitely less cumbersome to perform, it was shown to be less sensitive and more specific than its EIA companion (121, 122).

Detection of human antibodies against *Aspergillus* is also commercially available through different vendors. The Platelia™ Aspergillus IgG (Bio-Rad) is a two-phase indirect microplate immunoenzymatic technique used to detect IgG anti-*Aspergillus* antibodies in human serum or plasma. The VIRION/SERION ELISA classic *Aspergillus fumigatus* IgG/IgM/IgA (Serion Immundiagnostica & Institut Virion\Serion GmbH) is another quantitative alternative to detect antibodies against *A. fumigatus* in serum or plasma. In this case, separate detection of individual immunoglobulin classes offers a more detailed view for mycological monitoring of at-risk patients. Since an immune response is needed for antibodies to be present, this kind of assay is mostly used within the context of allergic or chronic *Aspergillus*-related pathologies (123, 124).

Two alternatives for *Aspergillus* detection, provided by IMMY, are available in Europe and USA. The first is based on the principle of Complement Fixation. As with other assays previously described, although a positive result strongly suggests *Aspergillus* infection or allergy, the organism must be repeatedly cultured or observed by microscopy for a definitive diagnosis. The second is an Immunodiffusion Test for aspergillosis, which is mostly restricted to diagnosis of allergic bronchopulmonary aspergillosis and aspergilloma.

5.4. Beta-D-Glucan Assays

Glucans are (1→3)-β-D-linked polymers of glucose that are part of the outer fungal cell wall. They can also be found in some bacteria, e.g., *Alcaligenes*, *Pseudomonas*, and *Streptococcus*, which produce glucan or glucan-like polymers, as well as in plant cellulose (125, 126). These polymers can be released or actively secreted from the cell wall as exopolymers and are well known to activate the immune system of mammals (127). Detection of β-D-glucan is commercially available through different manufacturers of assays based on the ability of β-D-glucan to react with *Limulus* Amebocyte Lysate (LAL). This compound, from horseshoe crab blood cells, was recently discovered to react with β-D-glucan (as well as endotoxin), by activation of factor G of the coagulation cascade (128).

The Fungitell™ assay (Associates of Cape Cod, Inc., East Falmouth, MA, USA) is CE-marked and FDA approved since 2004. In Japan, three companies propose assays for β-D-glucan detection. The Fungitec® G Test MK (Seikagaku Corporation, Tokyo, Japan) and B-G-Star B-D-glucan test (Maruha Nichiro Foods, Inc., Tokyo, Japan) are in vitro chromogenic endpoint quantitative assays, whereas the Wako test (Wako Pure Chemical Industries) is a kinetic turbidimetric test.

In summary, detection of circulating β-D-glucan seems to be a good indication to exclude invasive fungal infections, due to its high negative predictive value (112, 129, 130). General recommendation is to combine use of β-D-glucan with other surrogate markers such as mannan, GM, or fungal DNA, as in the majority of cases this combination not only provides species identification of the etiologic agent but also tends to increase β-D-glucan diagnostic accuracy (97). Assessment of the optimal sampling strategy for at-risk patients, the criteria to define a positive test result, the optimal cut-off value, and how to interpret the influence of concurrent bacteremia on diagnostic performance might be very helpful to determine optimal utilization of β-D-glucan testing for diagnosis of IFIs. Even though several commercially available solutions exist, the assay is still technically demanding and highly susceptible to external contamination during processing. Efforts need to be undertaken to simplify procedures and reduce the risk of false-positive results, although some of them are inherent to the almost omnipresence of glucans in the environment.

6. Molecular Biology-Based Methods

Recent progress in molecular biology, especially the arrival of real-time amplification and fluorescence-based detection tools, has allowed development and commercialization of different tests for detection and identification of fungi. Their development has often been linked to continuous efforts towards detection of pathogens associated with sepsis, a life-threatening condition associated with an altered inflammatory response to infection. Therefore, the vast majority of molecular biology-based tests commercially available for fungal detection are often coupled to that of bacteria and some antibiotic resistance markers. A summary of tests described in this section and their main characteristics is shown in Table 8. It is important to highlight that all these tests and kits are only available in the European market either for RUO or IVD purposes. The following sections will describe the tests available for sepsis as well as those tests available for specific detection of fungal species.

6.1. Sepsis-Based Assays

The LightCycler® SeptiFast Test M^{GRADE} (Roche Diagnostics GmbH, Mannheim, Germany) was the first test of this series of platforms dedicated to detection and identification of both bacteria and fungi directly from whole blood, as an aid in management of septic patients. This test is based on real-time PCR targeting species-specific ITS regions between the 18S and 5.8S rRNA genes. The assay can detect and identify five *Candida* species and *Aspergillus fumigatus*. As a constant for this type of assay, three main steps are applied from sample collection to final results, sample preparation, amplification, and real-time detection.

The initial sample for SeptiFast Test M^{GRADE} is whole ethylenediaminetetraacetic acid (EDTA) anticoagulated human blood, which is mechanically lysed using the SeptiFast Lyse Kit M^{GRADE} (Roche Diagnostics GmbH) and the MagNA Lyser® (Roche Diagnostics GmbH). While the method originally started from 3 ml of whole blood, the current version recommends starting with 1.5 ml. Only one-third of the total genomic DNA, purified by means of the SeptiFast Prep Kit M^{GRADE} (Roche Diagnostics

Table 8
Main features of the different molecular biology-based tests, commercially available, for fungal detection

Assay (manufacturer)	Species claimed	Sample type (volume)	Main features for: Sample preparation	PCR	Final identification
LightCycler® SeptiFast Test M^GRADE (Roche Diagnostics GmbH)	*Candida albicans*, *C. glabrata*, *C. tropicalis*, *C. krusei*, *C. parapsilosis*, and *Aspergillus fumigatus*	Whole blood (1.5 ml)	Isolation of total DNA Mostly manual[a]	Broad-range ITS PCR	Fluorescence real-time PCR signals and melting curves
Vyoo® Assay (SIRS-Lab GmbH)	Idem as above	Whole blood (up to 5 ml)	Selective enrichment of microbial DNA Automated process	Broad-range rDNA PCR	Fluorescence signals read on an automated array[b]
Magicplex™ Sepsis Real-Time Test (Seegene Inc.)	Idem as above	Whole blood	Selective human cell lysis and degradation of human DNA Manual or automated	Broad-range rDNA PCR	Fluorescence real-time PCR signals
SepsiTest™ Molzym GmbH & Co.KG)	Pan-yeast assay	Whole blood, heart valve tissue, and synovial fluid		Broad-range rDNA PCR	Sequencing of amplicons
Affigene® Aspergillus tracer (Cepheid AB)	*Aspergillus* spp.	Whole blood and plasma samples (0.5 ml)	Not specified[c]	Broad-range rDNA PCR	Fluorescence real-time PCR signal
MycAssay™ Aspergillus (Myconostica Ltd.)	*Aspergillus* spp.	Respiratory and serum samples	MycXtra DNA extraction for respiratory specimens Roche High Pure Template DNA kit for serum Manual processing	Broad-range 18S rDNA PCR	Fluorescence real-time PCR signal
MycAssay™ Pneumocystis (Myconostica Ltd.)	*Pneumocystis jirovecii*	Respiratory samples	MycXtra DNA extraction Manual processing	Mitochondrial large subunit PCR	Fluorescence real-time PCR signal

[a]Some published reports combine it with automated methods such as MagNA Pure® (Roche Diagnostics GmbH) or EasyMAG® (bioMérieux®)
[b]Former format used electrophoretic separation of amplicons
[c]Published report used Roche HighPure PCR Template Preparation Kit (Roche Diagnostics GmbH) (131)

GmbH), is dedicated to specific fungal detection. Amplification of targets occurs in the LightCycler® 2.0 instrument (Roche Diagnostics GmbH) and final interpretation of results is done by dedicated software (SIS SeptiFast Identification Software—Roche Diagnostics GmbH), which calculates melting point values and links the peak heights to a particular fungal species. The whole procedure is validated by reading positive signals coming from the internal control provided with the kit. Turnaround time ranges from 6 to 8 h.

The utility and performance of the LightCycler® SeptiFast Test M^{GRADE} in the clinical setting have been reported in more than 15 publications. A summary of results observed for detection of fungal pathogens is provided in Table 9. While blood culture is generally used as the reference for comparison, some authors prefer to evaluate performance through analysis of the overall sepsis or febrile episode and only a few studies have focused the analysis on how the test performed for detection of IFIs, based on published consensus definitions. To date, no clear evidence has been shown whether the LightCycler® SeptiFast Test M^{GRADE} can have an impact on management of patients with IFIs. None of the studies has tested serial specimens collected during the period at-risk, before blood cultures are drawn, and therefore it is still unclear whether fungal DNA can be considered or not as an early marker in development of invasive candidiasis (84). Nevertheless, results published thus far suggest that its use in combination with other methods (e.g., blood culture or GM detection) could be valuable. Whether LightCycler® SeptiFast Test M^{GRADE} has the potential to facilitate early evidence-based treatment decisions, antimicrobial selection, and adequacy of antimicrobial treatment, effectively designed observational or even interventional studies are needed to prove its clinical value in improving appropriate treatment and patient outcomes (142).

Another commercially available multiplex PCR sepsis-based test is the Vyoo® assay (SIRS-Lab GmbH, Jena, Germany). The Vyoo® assay can accommodate up to 5 ml EDTA-whole blood, hence increasing probability to detect low fungal burden. Furthermore, it uses a specific DNA-binding protein able to recognize unmethylated DNA motifs, much less frequent in human DNA than in bacteria, yeast, or molds (143). After a first step of mechanical lysis, microbial DNA is processed and enriched by affinity chromatography, which can significantly improve the signal-to-noise ratio between human and fungal DNA. The final detection of amplicons can be done either by electrophoretic separation or by hybridization methodologies (144). Available in Europe for diagnostic purposes since 2007, a new updated version was recently released including changes related to automated DNA isolation and enrichment as well as final readout of results which is now based on software-supported microarrays analysis.

Table 9
Summary of some published reports evaluating fungal detection performance of the LightCycler® SeptiFast Test M^{GRADE} (SF)

Population (# of patients)	Reference method (RM)	# of positive cases for the RM	# of positive cases for SF	Concordance (SF out of total by RM)	Comments	Reference
HU–FNP ($n=34$)	Blood culture	None	*A. fumigatus* ($n=2$)	*A. fumigatus* results linked to invasive infection confirmed by clinical data and mold isolation from BAL, 1 week later	Levels of galactomannan for both patients were continuously borderline	(134)
HU–FNP ($n=70$)	FNE	Candidemia ($n=0$) *A. fumigatus* ($n=12$)[a]	*Candida* ($n=3$) *A. fumigatus* ($n=5$)	*Candida* ($n=1$ of 3)[b] *A. fumigatus* ($n=5$ of 12)	*Aspergillus* infection link made on the basis of repetitive high levels of circulating GM	(133)
HU–FNP ($n=86$)	IFI definition[c] (3)	Probable aspergillosis ($n=2$) Possible aspergillosis ($n=1$) Possible candidiasis ($n=3$) Possible IFI ($n=1$)	*Candida* ($n=7$) *A. fumigatus* ($n=2$)	Probable aspergillosis (none of 2) Possible aspergillosis (1 of 1) Possible candidasis (3 of 3) Possible IFI (1 of 1)	Blood culture remained negative SeptiFast diagnosis preceded standard diagnosis in three cases	(135)
ICU & IC ($n=77$)	Blood culture	Candida ($n=4$)	*Candida* ($n=5$) *A. fumigatus* ($n=1$)	*Candida* (3 of 4)	Antifungal treatment against *Candida* could have been either adjusted earlier or implemented in one and three cases, respectively	(136)
ICU, ER, SW ($n=436$)	Sepsis episode	*Candida* ($n=9$)	*Candida* ($n=10$) *A. fumigatus* ($n=7$)	6//9 *Candida* (6 of 9) All 7 *A. fumigatus* PCR findings judged as workflow contamination	Study not designed for method comparison, not providing any conclusion on clinical utility of therapy adjustments	(132)
ICU, ER, GM ($n=200$)	Blood culture	*Candida* ($n=4$)	*Candida* ($n=4$) *A. fumigatus* ($n=0$)	*Candida* (2 of 4)		(137)

Not reported ($n=359$)	Blood culture	*Candida* ($n=10$)	*Candida* ($n=20$) *A. fumigatus* ($n=12$)	*Candida* (5 of 10) Two out of five patients linked to *A. fumigatus* infection	*A. fumigatus* PCR positive results in five patients For three of them, no test was performed to confirm *Aspergillus*	(138)
ICU & HU ($n=110$)	IFI definition (3)[c]	Probable aspergillosis ($n=3$) Possible aspergillosis ($n=4$) Candidemia ($n=0$) Candida colonization ($n=2$)	*Candida* (n=0) *A. fumigatus* (n=2)	Probable aspergillosis (2 of 3) Possible aspergillosis (none of 4) *Candida* colonization (none of 2)	No false-positive results for *Candida* spp. occurred, even in patients heavily colonized	(139)
HU, ICU ($n=86$)[d]	Blood culture	*Candida* ($n=1$)	*Candida* ($n=1$) *A. fumigatus* ($n=1$)	*Candida* (1 of 1) *Aspergillus* considered false-positive result[e]	Fungemia defined by microbial growth in one or more blood culture sets	(140)
ICU, SW, ER, OU, HU, NU, PU ($n=811$)	Blood culture	*Candida* ($n=16$)	*Candida* ($n=27$)	*Candida* ($n=13$ of 16)	No *Aspergillus* found Higher rate of positive results for patients undergoing empirical antimicrobial therapy	(141)

ICU intensive care unit, *IC* immunocompromised, *SW* surgical ward, *GM* general medicine, *HU* hematological unit, *OU* oncology unit, *PU* pediatrics unit, *NU* neonates unit, *FNP* febrile neutropenic patients, *IFI* invasive fungal infection, *FNE* febrile neutropenic episode

[a]Positive levels of galactomannan (>0.7 optical density index)

[b]Succesive positive PCR results

[c]Linked to high levels of procalcitonin

[d]33 febrile neutropenic patients

[e]Analytical galactomannan negative results on serum and no clinical evidence of aspergillosis

Furthermore, a positive control was included to guarantee complete validity of results. The total turnaround time is around 7 h.

The SepsiTest™ (Molzym GmbH & Co. KG, Bremen, Germany) is another bundled CE-IVD marked kit for the detection of bacteria, yeasts, and molds in whole blood and other body fluid samples. It uses a proprietary technology, which gets rid of human DNA through degradation of released DNA after selective lysis of human cells. Microorganisms are therefore enriched and subsequently lysed for final elution of their DNA. This manual processing can today be combined with selectNA™, an automated platform able to handle up to 12 samples, reducing hands-on time needed for sample preparation. The eluted DNA is then combined with broad-range rDNA PCR, which provides molecular detection of pathogens. SepsiTest™ supplies two separate assays with broad-range primers targeting eubacterial and fungal targets. The final identification needs sequencing of the amplification product, which is then compared to the highly variable region sequences contained in the online identification tool, provided by the manufacturer as well. The total turnaround time is around 8 h and the species currently identified include *C. albicans*, *C. parapsilosis*, *C. tropicalis*, *Cladosporium cladosporioides*, and *Malassezia* spp.

One of the latest CE-IVD marked sepsis-based kits is the Magicplex™ Sepsis Real-time Test (Seegene, Seoul, Korea). It claims detection of bacterial and fungal targets from EDTA-whole blood samples and is commercially available since November 2010. The same six fungal species as the LightCycler® SeptiFast Test M^{GRADE} are detected, by means of a broad-range PCR, compatible with different PCR platforms, including their own Seecycler™ (Seegene, Seoul, Korea). The company recently announced its partnership with Molzym GmbH & Co. KG to apply their microbial DNA enrichment technology and the selectNA™ automated platform to the assay process. At the moment this review was written, no study evaluating this test had been published.

6.2. Specific Assays

The MycAssay™ Aspergillus (Myconostica Ltd., Manchester, UK) is a CE-IVD marked, real-time PCR assay for the specific detection of *Aspergillus* DNA in serum and lower respiratory tract specimens. Myconostica Ltd., a diagnostic company spin-off from the University of Manchester, launched in 2010 this assay together with a product for extracting fungal DNA from human respiratory samples (MycXtra® DNA extraction).

The MycAssay™ Aspergillus PCR assay was validated for use with different platforms such as SmartCycler® (Cepheid, Sunnyvale, CA, USA) AB7500 (Life Technologies™, Carlsbad, CA, USA), LightCycler® 2.0, and Stratagene® M×3000™ (Agilent Technologies, Santa Clara, CA, USA). The gene targeted is the 18S rDNA and the analytical limit of detection (LoD) achieved is below 50 target copies. One interesting feature is that it provides a clinical cut-off for better interpretation of results in the clinical setting. The assay includes an internal amplification control in every reaction to highlight false-negative results. Due to its pan-*Aspergillus* capacity, the assay is presented in closed tube reactions and single-use reagents to reduce the risk of contamination.

Another PCR-based specific test for *Aspergillus* detection is available through Cepheid AB (Bromma, Sweden). The Affigene® Aspergillus tracer is a CE-IVD marked qualitative assay for detection of *Aspergillus* DNA directly from human whole blood and plasma samples. The kit provides reagents in a ready-to-use, single-use format, compatible with several real-time PCR platforms such as Mx3000P/3005P (Stratagene, La Jolla, USA), the iCycler iQ/iQ5 (BioRad Laboratories Inc, Hercules, CA, USA), the Rotor-Gene 3000 (Corbette Research, Sydney, Australia), and the ABI 7300 (Applied Biosystems, Foster City, USA). The analytical sensitivity in both plasma and whole blood is below 0.5 genome equivalents per microliter. The assay was evaluated in a population of neutropenic patients where two patients diagnosed as having proven or probable aspergillosis according to the EORTC/MSG criteria were the only ones showing a positive PCR result (145).

Myconostica Ltd., recently acquired by another English company (Lab 21 Ltd., Cambridge, UK), also commercializes the MycAssay™ Pneumocystis assay: the first commercially validated (CE-IVD marked) real-time PCR assay, for diagnosis of *Pneumocystis* infection in human clinical respiratory specimens, targeting the mitochondrial large subunit of *P. jirovecii*. It is recommended to be combined with MycXtra® DNA extraction and was validated for use on the same platforms as the MycAssay™ Aspergillus. Its LoD is below 35 target copies and an internal amplification control is used in every reaction. A recent multicenter study on 110 patients compared results to clinical diagnosis of *Pneumocystis* respiratory infection determined by non-molecular methods in accordance with local diagnostic procedures. Sensitivity and specificity were 93% and 90%, respectively, compared to values from Merifluor-*Pneumocystis* direct fluorescent antigen (146).

7. Unmet Needs and Trends for the Future

Diagnosis of fungal infections is still challenging and there is need for improvement. Efforts must continue towards development of less invasive, more standardized and reliable IVD tools, contributing to earlier and more accurate diagnosis of fungal infections, which will lead to more appropriate therapeutic decisions and improved outcomes of at-risk patients. Standardization and combination are two key concepts to keep in mind for successful development of future commercial tests, in addition to implementation of new technological advances in the field of diagnostics.

The standardization of techniques, especially those based on the detection of fungal nucleic acids is a must, and some international initiatives are already pursuing consensus guidelines for use of such techniques (147–149). This standardization is not only needed for the techniques but also for the design of evaluation studies. The heterogeneity of the different attained or at-risk populations hampers interpretation of results and the clinical utility of the evaluated techniques. Efforts must be made towards standardization of methodologies, which will indeed help enable formal validation of new tools, such as PCR-based ones (150).

Combination of information provided by different markers, or combination of distinct technologies, seems to be improving diagnostic accuracy. Examples have been provided regarding utility of merging information provided by detection of mannan and anti-mannan antibodies in the case of *Candida* infections (90), but this example has also been successfully implemented when combining information provided by detection of fungal antigens and nucleic acids by means of PCR-based assays (87, 139). Clinical guidelines also emphasize importance of combining information from different IVD tools (2).

Combining classical methodologies with the most recent technological advances is also how some manufacturers propose to move forward within the unmet diagnostic need. Probably the most relevant examples are those addressing positive blood cultures to rapidly and accurately identify the presence of fungal species—Yeast Traffic Light® PNA-FISH, Prove-it™ Fungi, and the BlackLight® Fungal ID kit.

Implementation of the newest and most recent technological advances will also warrant development of improved products. Appearance of technologies such as Digital PCR and better accessibility to next-generation sequencing technologies are examples of technologies that could arrive in the near future to the diagnostic field. Such technological jumps which occurred previously with serology were seen again with PCR and have recently occurred with the successful translation of mass spectrometry in the IVD field. MALDI-TOF mass spectrometry provides today an impressive way to rapidly identify fungal species from isolated colonies. Additional work is needed to better identify clinically relevant molds and since their ability to utilize carbohydrates and enzymatic substrates is quite diverse, the classical microbiological methods appear less promising than the potential of MALDI-TOF mass spectrometry. Future technological advances will make possible its use directly from positive blood culture bottles and maybe directly from the biological specimen.

Another example of technological combinations, even though not yet available for diagnostic purposes is the one provided by the Plex-ID (Abbot Wiesbaden, Germany). This system, based on the use of eubacterial and panfungal PCR, is able to detect nearly all known fungal targets by coupling genome-specific amplification and electrospray ionization mass spectrometry. It allows determination of the base composition from small quantities of nucleic acids in a complex mixture, which is then compared to the genomic information contained in a curated database, a key element of the system. The Ibis Bioscience database needs to be constantly updated with the latest information on newly identified microorganisms. The time to result is around 8 h and some papers have been published describing analytical performance and proof-of-principle on blind panels of bacterial isolates (151).

In addition to all facts previously mentioned, the future must also bring more automation. This would make tests easier and more standardized and in some cases bring them closer to the bedside, allowing reduced time to results, often crucial in some clinical situations (e.g., febrile neutropenia).

Acknowledgments

We would like to thank Karine Kaiser, Carole Vachon, and Ira Salkin for their helpful comments and critical review of this manuscript.

References

1. Fridkin SK (2005) The changing face of fungal infections in health care settings. Clin Infect Dis 41:1455–60
2. Walsh TJ, Anaissie EJ, Denning DW, Herbrecht R, Kontoyiannis DP, Marr KA, Morrison VA, Segal BH, Steinbach WJ, Stevens DA, van Burik JA, Wingard JR, Patterson TF (2008) Treatment of aspergillosis: clinical practice guidelines of the Infectious Diseases Society of America. Clin Infect Dis 46:327–360
3. Chandler FW, Kaplan W, Ajello L (1980) Color atlas and text of the histopathology of mycotic diseases. Year Book Medical Publishers, Chicago
4. Monheit JE, Cowan DF, Moore DG (1984) Rapid detection of fungi in tissue using calcofluor white and fluorescence microscopy. Arch Pathol Lab Med 108:616–618
5. Nolte FS, Williams JM, Jerris RC, Morello JA, Leitch CD, Matushek S, Schwabe LD, Dorigan F, Kocka FE (1993) Multicenter clinical evaluation of a continuous monitoring blood culture system using fluorescent-sensor technology (BACTEC 9240). J Clin Microbiol 31:552–557
6. Thorpe TC, Wilson ML, Turner JE, DiGuiseppi JL, Willert M, Mirrett S, Reller LB (1990) BacT/Alert: an automated colorimetric microbial detection system. J Clin Microbiol 28:1608–1612
7. Mirrett S, Hanson KE, Reller LB (2007) Controlled clinical comparison of VersaTREK and BacT/ALERT blood culture systems. J Clin Microbiol 45:299–302
8. Taschdjian CL, Burchall JJ, Kozinn PJ (1960) Rapid identification of *Candida albicans* by filamentation on serum and serum substitutes. Am J Dis Child 99:212–215
9. Crist AE Jr, Dietz TJ, Kampschroer K (1996) Comparison of the MUREX C. albicans, Albicans-Sure, and BactiCard Candida test kits with the germ tube test for presumptive

identification of *Candida albicans*. J Clin Microbiol 34:2616–2618
10. Heelan JS, Siliezar D, Coon K (1996) Comparison of rapid testing methods for enzyme production with the germ tube method for presumptive identification of *Candida albicans*. J Clin Microbiol 34:2847–2849
11. Carrillo-Muñoz AJ, Quindós G, Cárdenes CD, Alonso-Vargas R, Brió S, Arévalo P, Pemán J, Estivill D, Pontón J (2003) Performance of Bacticard Candida compared with the germ tube test for the presumptive identification of *Candida albicans*. Mycoses 46:467–470
12. Cárdenes CD, Carrillo-Muñoz AJ, Arias A, Rodríguez-Alvarez C, Torres-Lana A, Sierra A, Arévalo MP (2004) Comparative evaluation of four commercial tests for presumptive identification of *Candida albicans*. J Microbiol Methods 59:293–297
13. Fenn JP, Segal H, Blevins L, Fawson S, Newcomb-Gayman P, Carroll KC (1996) Comparison of the Murex *Candida albicans* CA50 test with germ tube production for identification of C. albicans. Diagn Microbiol Infect Dis 24:31–35
14. Baalham T, Brown A, Dunsmuir R, Shankland GS, Blackman M (2005) Comparative evaluation of a new biochemical test kit for the presumptive identification of *Candida albicans*. 15th European Congress of Clinical Microbiology and Infectious Diseases Copenhagen / Denmark. Abstract number: 1134_01_304 http://www.blackwellpublishing.com/eccmid15/abstract.asp?id=36567 Accessed 19 July 2011
15. Staib F (1962) *Cryptococcus neoformans* und *Guizotia abyssinica* (syn. G. oleifera D.C.). Farbreaktion für *Cr. neoformans*. Z Hyg Infektionskr Med Mikrobiol Immunol Virol 148:466–475
16. Denning DW, Stevens DA, Hamilton JR (1990) Comparison of *Guizotia abyssinica* seed extract (birdseed) agar with conventional media for selective identification of *Cryptococcus neoformans* in patients with acquired immunodeficiency syndrome. J Clin Microbiol 28:2565–2567
17. Fleming WH, Hopkins JM, Land GA (1977) New culture medium for the presumptive identificaion of *Candida albicans* and *Cryptococcus neoformans*. J Clin Microbiol 5: 236–243
18. Roberts GD, Horstmeier CD, Land GA, Foxworth JH (1978) Rapid urea broth test for yeasts. J Clin Microbiol 7:584–588
19. Zimmer BL, Roberts GD (1979) Rapid selective urease test for presumptive identification of *Cryptococcus neoformans*. J Clin Microbiol 10:380–381
20. Freydière AM, Robert R, Ploton C, Marot-Leblond A, Monerau F, Vandenesch F (2003) Rapid identification of *Candida glabrata* with a new commercial test, GLABRATA RTT. J Clin Microbiol 41:3861–3963
21. Freydière AM, Perry JD, Faure O, Willinger B, Tortorano AM, Nicholson A, Peman J, Verweij PE (2004) Routine use of a commercial test, GLABRATA RTT, for rapid identification of *Candida glabrata* in six laboratories. J Clin Microbiol 42:4870–4872
22. Pemán J, Aparisi N, García-Esteban C, Gobernado M (2004) Rapid identification of *Candida glabrata* using a new commercial kit. Rev Iberoam Micol 21:82–84
23. Willinger B, Wein S, Hirschl AM, Rotter ML, Manafi M (2005) Comparison of a new commercial test, GLABRATA RTT, with a dipstick test for rapid identification of *Candida glabrata*. J Clin Microbiol 43:499–501
24. Fenn JP, Billetdeaux E, Segal H, Skodack-Jones L, Padilla PE, Bale M, Carroll K (1999) Comparison of four methodologies for rapid and cost-effective identification of *Candida glabrata*. J Clin Microbiol 37: 3387–3389
25. Lopez J, Dalle F, Mantelin P, Moiroux P, Nierlich AC, Pacot A, Cuisenier B, Vagner O, Bonnin A (2001) Rapid identification of *Candida glabrata* based on trehalose and sucrose assimilation using Rosco diagnostic tablets. J Clin Microbiol 39:1172–1174
26. Freydière A-M, Parant F, Noel-Baron F, Crepy M, Treny A, Raberin H, Davidson A, Odds FC (2002) Identification of *Candida glabrata* by a 30-second trehalase test. J Clin Microbiol 40:3602–3605
27. Murray MP, Zinchuk R, Larone DH (2005) CHROMagar Candida as the sole primary medium for isolation of yeasts and as a source medium for the rapid-assimilation-of-trehalose test. J Clin Microbiol 43: 1210–1212
28. Odds FC, Hanson MF, Davidson AD, Jacobsen MD, Wright P, Whyte JA, Gow NA, Jones BL (2007) One year prospective survey of *Candida* bloodstream infections in Scotland. J Med Microbiol 56:1066–1075
29. Orenga S, James AL, Manafi M, Perry JD, Pincus DH (2009) Enzymatic substrates in microbiology. J Microbiol Methods 79: 139–155

30. Perry JD, Freydière AM (2007) The application of chromogenic media in clinical microbiology. J Appl Microbiol 103:2046–2055
31. Ghelardi E, Pichierri G, Castagna B, Barnini S, Tavanti A, Campa M (2008) Efficacy of Chromogenic Candida Agar for isolation and presumptive identification of pathogenic yeast species. Clin Microbiol Infect 14: 141–147
32. Sendid B, François N, Standaert A, Dehecq E, Zerimech F, Camus D, Poulain D (2007) Prospective evaluation of the new chromogenic medium CandiSelect 4 for differentiation and presumptive identification of the major pathogenic *Candida* species. J Med Microbiol 56:495–499
33. Odds FC, Bernaerts R (1994) CHROMagar Candida, a new differential isolation medium for presumptive identification of clinically important *Candida* species. J Clin Microbiol 32:1923–1929
34. Powell HL, Sand CA, Rennie RP (1998) Evaluation of CHROMagar Candida for presumptive identification of clinically important *Candida* species. Diagn Microbiol Infect Dis 32:201–204
35. Baumgartner C, Freydière AM, Gille Y (1996) Direct identification and recognition of yeast species from clinical material by using albicans ID and CHROMagar Candida plates. J Clin Microbiol 34:454–456
36. Fricker-Hidalgo H, Vandapel O, Duchesne MA, Mazoyer MA, Monget D, Lardy B, Lebeau B, Freney J, Ambroise-Thomas P, Grillot R (1996) Comparison of the new API Candida system to the ID 32C system for identification of clinically important yeast species. J Clin Microbiol 34:1846–1848
37. Bernal S, Martín Mazuelos E, Chávez M, Coronilla J, Valverde A (1998) Evaluation of the new API Candida system for identification of the most clinically important yeast species. Diagn Microbiol Infect Dis 32:217–221
38. Buchaille L, Freydière AM, Guinet R, Gille Y (1998) Evaluation of six commercial systems for identification of medically important yeasts. Eur J Clin Microbiol Infect Dis 17:479–488
39. Campbell CK, Davey KG, Holmes AD, Szekely A, Warnock DW (1999) Comparison of the API Candida system with the AUXACOLOR system for identification of common yeast pathogens. J Clin Microbiol 37:821–823
40. Paugam A, Benchetrit M, Fiacre A, Tourte-Schaefer C, Dupouy-Camet J (1999) Comparison of four commercialized biochemical systems for clinical yeast identification by colour-producing reactions. Med Mycol 37:11–17
41. Buesching WJ, Kurek K, Roberts GD (1979) Evaluation of the modified API 20C system for identification of clinically important yeasts. J Clin Microbiol 9:565–569
42. Land GA, Harrison BA, Hulme KL, Cooper BH, Byrd JC (1979) Evaluation of the new API 20C strip for yeast identification against a conventional method. J Clin Microbiol 10:357–364
43. Bergan T, Hollum AB, Vangdal M (1982) Evaluation of four commercial biochemical test systems for identification of yeasts. Eur J Clin Microbiol 1:217–222
44. Schuffenecker I, Freydière A, de Montclos H, Gille Y (1993) Evaluation of four commercial systems for identification of medically important yeasts. Eur J Clin Microbiol Infect Dis 12:255–260
45. Willemsen M, Breynaert J, Lauwers S (1997) Comparison of Auxacolor with API 20 C Aux in yeast identification. Clin Microbiol Infect 3:369–375
46. Ramani R, Gromadzki S, Pincus DH, Salkin IF, Chaturvedi V (1998) Efficacy of API 20C and ID 32C systems for identification of common and rare clinical yeast isolates. J Clin Microbiol 36:3396–3398
47. Sand C, Rennie RP (1999) Comparison of three commercial systems for the identification of germ-tube negative yeast species isolated from clinical specimens. Diagn Microbiol Infect Dis 33:223–229
48. Wadlin JK, Hanko G, Stewart R, Pape J, Nachamkin I (1999) Comparison of three commercial systems for identification of yeasts commonly isolated in the clinical microbiology laboratory. J Clin Mcrobiol 37:1967–1970
49. Gündes SG, Gulenc S, Bingol R (2001) Comparative performance of Fungichrom I, Candifast and API 20C Aux systems in the identification of clinically significant yeasts. J Med Microbiol 50:1105–1110
50. Milan EP, Malheiros ES, Fischman O, Colombo AL (1997) Evaluation of the AUXACOLOR system for the identification of clinical yeast isolates. Mycopathologia 137:153–157
51. Sheppard DC, deSouza E, Hashmi Z, Robson HG, René P (1998) Evaluation of the Auxacolor system for biochemical identification of medically important yeasts. J Clin Microbiol 36:3726–3727
52. Romney MG, Bryce EA, Rennie RP, Sand CA (2000) Rapid identification of clinical yeast isolates using the colorimetric

AUXACOLOR system. Diagn Microbiol Infect Dis 36:137–138
53. Latouche GN, Daniel HM, Lee OC, Mitchell TG, Sorrell TC, Meyer W (1997) Comparison of use of phenotypic and genotypic characteristics for identification of species of the anamorph genus *Candida* and related teleomorph yeast species. J Clin Microbiol 35:3171–3180
54. Marklein G, Josten M, Klanke U, Müller E, Horré R, Maier T, Wenzel T, Kostrzewa M, Bierbaum G, Hoerauf A, Sahl HG (2009) Matrix-assisted laser desorption ionization-time of flight mass spectrometry for fast and reliable identification of clinical yeast isolates. J Clin Microbiol 47:2912–2917
55. Kitch TT, Jacobs MR, McGinnis MR, Appelbaum PC (1996) Ability of RapID Yeast Plus System to identify 304 clinically significant yeasts within 5 hours. J Clin Microbiol 34:1069–1071
56. Espinel-Ingroff A, Stockman L, Roberts G, Pincus D, Pollack J, Marler J (1998) Comparison of RapID yeast plus system with API 20C system for identification of common, new, and emerging yeast pathogens. J Clin Microbiol 36:883–886
57. Heelan JS, Sotomayor E, Coon K, D'Arezzo JB (1998) Comparison of the rapid yeast plus panel with the API20C yeast system for identification of clinically significant isolates of Candida species. J Clin Microbiol 36: 1443–1445
58. Sanguinetti M, Porta R, Sali M, La Sorda M, Pecorini G, Fadda G, Posteraro B (2007) Evaluation of VITEK 2 and RapID Yeast Plus systems for yeast species identification: experience at a large clinical microbiology laboratory. J Clin Microbiol 45:1343–1346
59. Bowman PI, Ahearn DG (1975) Evaluation of the Uni-Yeast-Tek kit for the identification of medically important yeasts. J Clin Microbiol 2:354–358
60. Bowman PI, Ahearn DG (1976) Evaluation of commercial systems for the identification of clinical yeast isolates. J Clin Microbiol 4:49–53
61. Cooper BH, Johnson JB, Thaxton ES (1978) Clinical evaluation of the Uni-Yeast-Tek system for rapid presumptive identification of medically important yeasts. J Clin Microbiol 7:349–355
62. Salkin IF, Land GA, Hurd NJ, Goldson PR, McGinnis MR (1987) Evaluation of YeastIdent and Uni-Yeast-Tek yeast identification systems. J Clin Microbiol 25: 624–627
63. McGinnis MR, Molina TC, Pierson DL, Mishra SK (1996) Evaluation of the Biolog MicroStation system for yeast identification. J Med Vet Mycol 34:349–352
64. Land GA, Salkin IF, El-Zaatari M, McGinnis MR, Hashem G (1991) Evaluation of the Baxter-MicroScan 4-hour enzyme-based yeast identification system. J Clin Microbiol 29: 718–722
65. St Germain G, Beauchesne D (1991) Evaluation of the MicroScan rapid yeast identification panel. J Clin Microbiol 29:2296–2299
66. Riddle DL, Giger O, Miller L, Hall GS, Woods GL (1994) Clinical comparison of the Baxter MicroScan yeast identification panel and the Vitek Yeast Biochemical Card. Am J Clin Pathol 101:438–442
67. Crist AE Jr, Johnson LM, Burke PJ (1996) Evaluation of the Microbial Identification System for identification of clinically isolated yeasts. J Clin Microbiol 34:2408–2410
68. Kellogg JA, Bankert DA, Chaturvedi V (1998) Limitations of the current Microbial Identification System for identification of clinical yeast isolates. J Clin Microbiol 36: 1197–1200
69. Kellogg JA, Bankert DA, Chaturvedi V (1999) Variation in Microbial Identification System accuracy for yeast identification depending on commercial source of Sabouraud dextrose agar. J Clin Microbiol 37: 2080–2083
70. Oblack DL, Rhodes JC, Martin WJ (1981) Clinical evaluation of the AutoMicrobic system Yeast Biochemical Card for rapid identification of medically important yeasts. J Clin Microbiol 13:351–355
71. Land G, Stotler R, Land K, Staneck J (1984) Update and evaluation of the AutoMicrobic yeast identification system. J Clin Microbiol 20:649–652
72. el-Zaatari M, Pasarell L, McGinnis MR, Buckner J, Land GA, Salkin IF (1990) Evaluation of the updated Vitek yeast identification data base. J Clin Microbiol 28: 1938–1941
73. Fenn JP, Segal H, Barland B, Denton D, Whisenant J, Chun H, Christofferson K, Hamilton L, Carroll K (1994) Comparison of updated Vitek Yeast Biochemical Card and API 20C yeast identification systems. J Clin Microbiol 32:1184–1187
74. Dooley DP, Beckius ML, Jeffrey BS (1994) Misidentification of clinical yeast isolates by using the updated Vitek Yeast Biochemical Card. J Clin Microbiol 32:2889–2892
75. Aubertine CL, Rivera M, Rohan SM, Larone DH (2006) Comparative study of the new colorimetric VITEK 2 yeast identification card fluorometric card and of CHROMagar

Candida as a source medium with the new card. J Clin Microbiol 44:227–228

76. Loïez C, Wallet F, Sendid B, Courcol RJ (2006) Evaluation of VITEK 2 colorimetric cards versus fluorimetric cards for identification of yeasts. Diagn Microbiol Infect Dis 56: 455–457

77. Hata DJ, Hall L, Fothergill AW, Larone DH, Wengenack NL (2007) Multicenter evaluation of the new VITEK 2 advanced colorimetric yeast identification versus the older card. J Clin Microbiol 45:1087–1092

78. Valenza G, Strasen J, Schafer F, Frosch M, Kurzai O, Abele-Horn M (2008) Evaluation of new colorimetric Vitek 2 yeast identification card by use of different source media. J Clin Microbiol 46:3784–3787

79. Vijgen S, Nys S, Naesens R, Magerman K, Boel A, Cartuyvels R (2011) Comparison of Vitek identification and antifungal susceptibility testing methods to DNA sequencing and Sensititre YeastOne antifungal testing. Med Mycol 49:107–110

80. Hasyn JJ, Buckley HR (1982) Evaluation of the AutoMicrobic system for identification of yeasts. J Clin Microbiol 16:901–904

81. Pfaller MA, Preston T, Bale M, Koontz FP, Body BA (1988) Comparison of the Quantum II, API Yeast Ident, and AutoMicrobic systems for identification of clinical yeast isolates. J Clin Microbiol 26:2054–2058

82. Quiles-Melero I, García-Rodriguez J, Romero-Gómez MP, Gómez-Sánchez P, Mingorance J (2011) Rapid identification of yeasts from positive blood culture bottles by pyrosequencing. Eur J Clin Microbiol Infect Dis 30:21–24

83. Fukazawa Y (1989) Antigenic structure of *Candida albicans*. Immunochemical basis of the serologic specificity of the mannans in yeasts. Immunol Ser 47:37–62

84. Bille J (2010) New nonculture-based methods for the diagnosis of invasive candidiasis. Curr Opin Crit Care 16:460–464

85. Oliveri S, Trovato L, Betta P, Romeo MG, Nicoletti G (2008) Experience with the Platelia Candida ELISA for the diagnosis of invasive candidosis in neonatal patients. Clin Microbiol Infect 14:391–393

86. Platelia™ Candida AB/AC/AK (2009) Bio-Rad instructions for use. <http://www3.bio-rad.com/cmc_upload/Literature/inserts/62799_881046_GB.pdf> Accessed 01 May 2011

87. Alam FF, Mustafa AS, Khan ZU (2007) Comparative evaluation of (1, 3)-β-D-glucan, mannan and anti-mannan antibodies, and Candida species-specific snPCR in patients with candidemia. BMC Infect Dis 7:103–111

88. Arendrup MC, Bergmann OJ, Larsson L, Nielsen HV, Jarløv JO, Christensson B (2010) Detection of candidaemia in patients with and without underlying hematological disease. Clin Microbiol Infect 16:855–862

89. Sendid B, Caillot D, Baccouch-Humbert B, Klingspor L, Grandjean M, Bonnin A, Poulain D (2003) Contribution of the Platelia Candida-specific antibody and antigen tests to early diagnosis of systemic *Candida tropicalis* infection in neutropenic adults. J Clin Microbiol 41:4551–4558

90. Mikulska M, Calandra T, Sanguinetti M, Poulain D, Viscoli C, Third European Conference on Infections in Leukemia Group (2010) The use of mannan antigen and anti-mannan antibodies in the diagnosis of invasive candidiasis: recommendations from the Third European Conference on Infections in Leukemia. Crit Care 14:R222

91. Downey LC, Smith PB, Benjamin DK, Cohen-Wolkowiez M (2010) Recent advances in the detection of neonatal Candidiasis. Curr Fungal Infect Rep 4:17–22

92. Lunel FM, Mennink-Kersten MA, Ruegebrink D, van der Lee HA, Donnelly JP, Blijlevens NM, Verweij PE (2009) Value of Candida serum markers in patients with invasive candidiasis after myeloablative chemotherapy. Diagn Microbiol Infect Dis 64:408–15

93. Wulf GG, Conradi I, Rüchel R (2004) Diagnostic value of antibody detection against *Candida* germ tube antigens in a patient with hepatosplenic candidosis. Mycoses 47(Suppl 1):11–14

94. Mitsutake K, Miyazaki T, Tashiro T, Yamamoto Y, Kakeya H, Otsubo T, Kawamura S, Hossain MA, Noda T, Hirakata Y, Kohno S (1996) Enolase antigen, mannan antigen, Cand-Tec antigen, and β-glucan in patients with candidemia. J Clin Microbiol 34: 1918–1921

95. Mori T, Matsumura M (1999) Clinical evaluation of diagnostic methods using plasma and/or serum for three mycoses: aspergillosis, candidosis, and pneumocystosis. Nippon Ishinkin Gakkai Zasshi 40:223–230

96. Bär W, Hecker H (2002) Diagnosis of systemic *Candida* infections in patients of the intensive care unit. Significance of serum antigens and antibodies. Mycoses 45:22–28

97. Pazos C, Moragues MD, Quindós G, Pontón J, Palacio A (2006) Diagnostic potential of $(1 \rightarrow 3)$- β-D-glucan and anti-*Candida albicans* germ tube antibodies for the diagnosis and therapeutic monitoring of invasive candidiasis in neutropenic adult patients. Rev Iberoam Micol 23:209–215

98. Moragues MD, Ortiz N, Iruretagoyena JR, García-Ruiz JC, Amutio E, Rojas A, Mendoza J, Quindós G, Pontón-San Emeterio J (2004) Evaluation of a new commercial test (*Candida albicans* IFA IgG) for the serodiagnosis of invasive candidiasis. Enferm Infecc Microbiol Clin 22:83–88
99. Pemán J, Zaragoza R, Quindós G, Alkorta M, Cuétara MS, Camarena JJ, Ramírez P, Giménez MJ, Martín-Mazuelos E, Linares-Sicilia MJ, Pontón J, The study group Candida albicans Germ Tube Antibody Detection in Critically Ill Patients (CAGTAUCI) (2011) Clinical factors associated with a *Candida albicans* germ tube antibody positive test in intensive care unit patients. BMC Infect Dis 11:60–65
100. Zaragoza R, Pemán J, Quindós G, Iruretagoyena JR, Cuétara MS, Ramírez P, Gómez MD, Camarena JJ, Viudes A, Pontón J, On behalf of the study group Candida albicans Germ Tube Antibody Detection in Critically Ill Patients (CAGTAUCI) (2009) Clinical significance of the detection of *Candida albicans* germ tube-specific antibodies in critically ill patients. Clin Microbiol Infect 15:592–595
101. Temstet A, Roux P, Poirot JL, Ronin O, Dromer F (1992) Evaluation of a monoclonal antibody-based latex agglutination test for diagnosis of cryptococcosis: comparison with two tests using polyclonal antibodies. J Clin Microbiol 30:2544–2550
102. Tanner DC, Weinstein MP, Fedorciw B, Joho KL, Thorpe JJ, Reller L (1994) Comparison of commercial kits for detection of cryptococcal antigen. J Clin Microbiol 32:1680–1684
103. Blevins LB, Fenn J, Segal H, Newcomb-Gayman P, Carroll KC (1995) False-positive cryptococcal antigen latex agglutination caused by disinfectants and soaps. J Clin Microbiol 33:1674–1675
104. Wilson DA, Sholtis M, Parshall S, Hall GS, Procop GW (2011) False-positive cryptococcal antigen test associated with use of BBL Port-A-Cul transport vials. J Clin Microbiol 49:702–703
105. Heelan JS, Corpus L, Kessimian N (1991) False-positive reactions in the latex agglutination test for *Cryptococcus neoformans* antigen. J Clin Microbiol 29:1260–1261
106. McFadden DC, Zaragoza O, Casadevall A (2004) Immunoreactivity of cryptococcal antigen is not stable under prolonged incubations in human serum. J Clin Microbiol 42:2786–2788
107. Saha DC, Xess I, Biswas A, Bhowmik DM, Padma MV (2009) Detection of *Cryptococcus* by conventional, serological and molecular methods. J Med Microbiol 58:1098–1105
108. Jaye DL, Waites KB, Parker B, Bragg SL, Moser SA (1998) Comparison of two rapid latex agglutination tests for detection of cryptococcal capsular polysaccharide. Am J Clin Pathol 109:634–641
109. Babady NE, Bestrom JE, Jespersen DJ, Jones MF, Beito EM, Binnicker MJ, Wengenack NL (2009) Evaluation of three commercial latex agglutination kits and a commercial enzyme immunoassay for the detection of cryptococcal antigen. Med Mycol 47:336–338
110. Mennink-Kersten MA, Donnelly JP, Verweij PE (2004) Detection of circulating galactomannan for the diagnosis and management of invasive aspergillosis. Lancet Infect Dis 4:349–57
111. Verdaguer V, Walsh TJ, Hope W, Cortez KJ (2007) Galactomannan antigen detection in the diagnosis of invasive aspergillosis. Expert Rev Mol Diagn 7:21–32
112. Thornton CR (2010) Detection of invasive aspergillosis. Adv Appl Microbiol 70:187–216
113. Chen SC, Kontoyiannis DP (2010) New molecular and surrogate biomarker-based tests in the diagnosis of bacterial and fungal infection in febrile neutropenic patients. Curr Opin Infect Dis 23:567–77
114. Klont RR, Mennink-Kersten MA, Verweij PE (2004) Utility of *Aspergillus* antigen detection in specimens other than serum specimens. Clin Infect Dis 39:1467–1474
115. Husain S, Clancy CJ, Nguyen MH, Swartzentruber S, Leather H, LeMonte AM, Durkin MM, Knox KS, Hage CA, Bentsen C, Singh N, Wingard JR, Wheat LJ (2008) Performance characteristics of the Platelia *Aspergillus* enzyme immunoassay for detection of Aspergillus galactomannan antigen in bronchoalveolar lavage fluid. Clin Vaccine Immunol 15:1760–1763
116. Hage CA, Knox KS, Davis TE, Wheat LJ (2011) Antigen detection in bronchoalveolar lavage fluid for diagnosis of fungal pneumonia. Curr Opin Pulm Med 17:167–171
117. Maertens J, Maertens V, Theunissen K, Meersseman W, Meersseman P, Meers S, Verbeken E, Verhoef G, Van Eldere J, Lagrou K (2009) Bronchoalveolar lavage fluid galactomannan for the diagnosis of invasive pulmonary aspergillosis in patients with hematologic diseases. Clin Infect Dis 49: 1688–1693
118. Hsu LY, Ding Y, Phua J, Koh LP, Chan DS, Khoo KL, Tambyah PA (2010) Galactomannan testing of bronchoalveolar lavage fluid is use-

ful for diagnosis of invasive pulmonary aspergillosis in hematology patients. BMC Infect Dis 10:44–49

119. Nguyen MH, Jaber R, Leather HL, Wingard JR, Staley B, Wheat LJ, Cline CL, Baz M, Rand KH, Clancy CJ (2007) Use of bronchoalveolar lavage to detect galactomannan for diagnosis of pulmonary aspergillosis among nonimmunocompromised hosts. J Clin Microbiol 45:2787–2792
120. Steinbach WJ (2010) Invasive aspergillosis in pediatric patients. Curr Med Res Opin 26: 1779–1787
121. Verweij PE, Stynen D, Rijs AJ, de Pauw BE, Hoogkamp-Korstanje JA, Meis JF (1995) Sandwich enzyme-linked immunosorbent assay compared with Pastorex latex agglutination test for diagnosing invasive aspergillosis in immunocompromised patients. J Clin Microbiol 33:1912–1914
122. Machetti M, Feasi M, Mordini N, Van Lint MT, Bacigalupo A, Latgé JP, Sarfati J, Viscoli C (1998) Comparison of an enzyme immunoassay and a latex agglutination system for the diagnosis of invasive aspergillosis in bone marrow transplant recipients. Bone Marrow Transplant 21:917–921
123. Fricker-Hidalgo H, Coltey B, Llerena C, Renversez JC, Grillot R, Pin I, Pelloux H, Pinel C (2010) Recombinant allergens combined with biological markers in the diagnosis of allergic bronchopulmonary aspergillosis in cystic fibrosis patients. Clin Vaccine Immunol 17:1330–1336
124. Tsagarakis N, Kentrou N, Margariti V, Malgarinou A, Maurea S, Tsagarakis I, Anastasakou E (2009) Anti-aspergillus antibodies and galactomannan antigen detection for serodiagnosis of aspergillosis. ECCMID Meeting. Abstract number: P905
125. Stone BA, Clarke AE (1992) Chemistry and biology of (1→3)-β-glucans. La Trobe University Press, Melbourne, Australia
126. Mennink-Kersten MA, Ruegebrink D, Verweij PE (2008) *Pseudomonas aeruginosa* as a cause of 1,3-β-D-glucan assay reactivity. Clin Infect Dis 46:1930–1931
127. Suzuki T, Ohno N, Saito K, Yadomae T (1992) Activation of the complement system by (1→3)-beta-D-glucans having different degrees of branching and different ultrastructures. J Pharmacobiodyn 15:277–285
128. Ohki M, Nakamura T, Morita T, Iwanaga S (1980) A new endotoxin sensitive factor associated with hemolymph coagulation system of horseshoe crab (Limulidae). FEBS Lett 120:217–20
129. Marty FM, Koo S (2009) Role of (1→3)-β-D-glucan in the diagnosis of invasive aspergillosis. Med Mycol 47(Suppl 1):S233–240
130. Maertens J, Meersseman W, Van Bleyenbergh P (2009) New therapies for fungal pneumonia. Curr Opin Infect Dis 22:183–190
131. Torelli R, Posteraro B, De Carolis E, De Pascale G, Caira M, Bello G, Antonelli M, Pagano L, Fadda G, Sanguinetti M (2011) Comparison of MycAssay Aspergillus test with galactomannan detection in bronchoalveolar lavage (BAL) fluid samples of hematological and ICU patients. Poster 1941. 21st European society of clinical microbiology and infectious diseases meeting, Milan, Italy
132. Lehmann LE, Alvarez J, Hunfeld KP, Goglio A, Kost GJ, Louie RF, Raglio A, Regueiro BJ, Wissing H, Stüber F (2009) Potential clinical utility of polymerase chain reaction in microbiological testing for sepsis. Crit Care Med 37:3085–3090
133. von Lilienfeld-Toal M, Lehmann LE, Raadts AD, Hahn-Ast C, Orlopp KS, Marklein G, Purr I, Cook G, Hoeft A, Glasmacher A, Stüber F (2009) Utility of a commercially available multiplex real-time PCR assay to detect bacterial and fungal pathogens in febrile neutropenia. J Clin Microbiol 47:2405–2410
134. Mancini N, Clerici D, Diotti R, Perotti M, Ghidoli N, De Marco D, Pizzorno B, Emrich T, Burioni R, Ciceri F, Clementi M (2008) Molecular diagnosis of sepsis in neutropenic patients with hematological malignancies. J Med Microbiol 57:601–604
135. Lamoth F, Jaton K, Prod'hom G, Senn L, Bille J, Calandra T, Marchetti O (2010) Multiplex blood PCR in combination with blood cultures for improvement of microbiological documentation of infection in febrile neutropenia. J Clin Microbiol 48:3510–3516
136. Dierkes C, Ehrenstein B, Siebig S, Linde HJ, Reischl U, Salzberger B (2009) Clinical impact of a commercially available multiplex PCR system for rapid detection of pathogens in patients with presumed sepsis. BMC Infect Dis 9:126–132
137. Louie RF, Tang Z, Albertson TE, Cohen S, Tran NK, Kost GJ (2008) Multiplex polymerase chain reaction detection enhancement of bacteremia and fungemia. Crit Care Med 36:1487–92
138. Westh H, Lisby G, Breysse F, Böddinghaus B, Chomarat M, Gant V, Goglio A, Raglio A, Schuster H, Stuber F, Wissing H, Hoeft A (2009) Multiplex real-time PCR and blood culture for identification of bloodstream

pathogens in patients with suspected sepsis. Clin Microbiol Infect 15:544–51

139. Maubon D, Hamidfar-Roy R, Courby S, Vesin A, Maurin M, Pavese P, Ravanel N, Bulabois CE, Brion JP, Pelloux H, Timsit JF (2010) Therapeutic impact and diagnostic performance of multiplex PCR in patients with malignancies and suspected sepsis. J Infect 61:335–42

140. Yanagihara K, Kitagawa Y, Tomonaga M, Tsukasaki K, Kohno S, Seki M, Sugimoto H, Shimazu T, Tasaki O, Matsushima A, Ikeda Y, Okamoto S, Aikawa N, Hori S, Obara H, Ishizaka A, Hasegawa N, Takeda J, Kamihira S, Sugahara K, Asari S, Murata M, Kobayashi Y, Ginba H, Sumiyama Y, Kitajima M (2010) Evaluation of pathogen detection from clinical samples by real-time polymerase chain reaction using a sepsis pathogen DNA detection kit. Crit Care 14:R159

141. Sachse S, Straube E, Lehmann M, Bauer M, Rußwurm S, Schmidt K-H (2009) Truncated hCGBP improves NAT-based detection of bacterial species in human samples. J Clin Microbiol 47:1050–1057

142. Kortgen A, Bloos F, Sachse S, Lehmann M, Straube E, Reinhart K, Bauer M (2009) Clinical relevance of fungal DNAemia detected by multiplex PCR in septic patients. 4th international congress of the german sepsis society, Weimar, Germany.

143. An assessment of the potential clinical utility of a new multiplex-PCR assay (Vyoo®) in the management of ICU patients with sepsis. http://apps.who.int/trialsearch/Trial.aspx?TrialID=DRKS00000612 Accessed 19 July 2011

144. Wellinghausen N, Kochem AJ, Disqué C, Mühl H, Gebert S, Winter J, Matten J, Sakka SG (2009) Diagnosis of bacteremia in whole-blood samples by use of a commercial universal 16S rRNA gene-based PCR and sequence analysis. J Clin Microbiol 47:2759–65

145. Westh H, Lisby G, Breysse F, Böddinghaus B, Chomarat M, Gant V, Goglio A, Raglio A, Schuster H, Stuber F, Wissing H, Hoeft A (2009) Multiplex real-time PCR and blood culture for identification of bloodstream pathogens in patients with suspected sepsis. Clin Microbiol Infect 15:544–51

146. Maubon D, Hamidfar-Roy R, Courby S, Vesin A, Maurin M, Pavese P, Ravanel N, Bulabois CE, Brion JP, Pelloux H, Timsit JF (2010) Therapeutic impact and diagnostic performance of multiplex PCR in patients with malignancies and suspected sepsis. J Infect 61:335–42

147. Hauser PM, Bille J, Lass-Flörl C, Geltner C, Feldmesser M, Levi M, Patel H, Muggia V, Alexander B, Hughes M, Follett SA, Cui X, Leung F, Morgan G, Moody A, Perlin DS, Denning DW (2011) Multicenter, prospective clinical evaluation of respiratory samples from subjects at risk for Pneumocystis jirovecii infection by use of a commercial real-time PCR assay. J Clin Microbiol 49:1872–8

148. White PL, Perry MD, Loeffler J, Melchers W, Klingspor L, Bretagne S, McCulloch E, Cuenca-Estrella M, Finnstrom N, Donnelly JP, Barnes RA, European Aspergillus PCR Initiative (2010) Critical stages of extracting DNA from Aspergillus fumigatus in whole-blood specimens. J Clin Microbiol. 48:3753–3755

149. Klingspor L, Loeffler J (2009) Aspergillus PCR formidable challenges and progress. Med Mycol 47(Suppl 1):S241–7

150. Mengoli C, Cruciani M, Barnes RA, Loeffler J, Donnelly JP (2009) Use of PCR for diagnosis of invasive aspergillosis: systematic review and meta-analysis. Lancet Infect Dis 9:89–96

151. Baldwin CD, Howe GB, Sampath R, Blyn LB, Matthews H, Harpin V, Hall TA, Drader JJ, Hofstadler SA, Eshoo MW, Rudnick K, Studarus K, Moore D, Abbott S, Janda JM, Whitehouse CA (2009) Usefulness of multilocus polymerase chain reaction followed by electrospray ionization mass spectrometry to identify a diverse panel of bacterial isolates. Diagn Microbiol Infect Dis 63:403–408

Chapter 3

Rapid Methods for the Extraction and Archiving of Molecular Grade Fungal Genomic DNA

Andrew M. Borman, Michael Palmer, and Elizabeth M. Johnson

Abstract

The rapid and inexpensive extraction of fungal genomic DNA that is of sufficient quality for molecular approaches is central to the molecular identification, epidemiological analysis, taxonomy, and strain typing of pathogenic fungi. Although many commercially available and in-house extraction procedures do eliminate the majority of contaminants that commonly inhibit molecular approaches, the inherent difficulties in breaking fungal cell walls lead to protocols that are labor intensive and that routinely take several hours to complete. Here we describe several methods that we have developed in our laboratory that allow the extremely rapid and inexpensive preparation of fungal genomic DNA.

Key words: Genomic DNA, PCR, FTA matrix, Yeast, Mold, Fungi

1. Introduction

Molecular identification approaches play an increasingly pivotal role in medical and veterinary mycology, at least in part due to the rapidly increasing number of fungal species associated with human infections (1–4), the description of cryptic fungal species in many pathogenic species (5–8), and the demonstration of species-specific differences in antifungal susceptibility profiles for key species of both yeasts and molds (9–13). For most fungi, the presence of significant quantities of PCR inhibitors in fungal cultures and difficulties inherent in breaking fungal cell walls (14) demand prior purification of total fungal genomic DNA for most molecular approaches. Existing protocols range from simple boiling for the rapid preparation of minute quantities of relatively impure DNA to more elaborate methodologies involving physical disruption of

Louise O'Connor and Barry Glynn (eds.), *Fungal Diagnostics: Methods and Protocols*, Methods in Molecular Biology, vol. 968, DOI 10.1007/978-1-62703-257-5_3, © Springer Science+Business Media New York 2013

fungal elements coupled with purification steps using column-based technologies or extractions with hazardous chemicals or solvents, and routinely cannot be accomplished in less than 1 h (15–24).

Whatman FTA filter matrices (GE Healthcare) are fibrous cards containing chelators and denaturants designed to lyse and inactivate most microorganisms, releasing large nucleic acids that subsequently become physically and stably trapped within the fibers of the FTA card. FTA filter cards have been successfully applied to the extraction and storage of DNA from viruses, bacteria, plants, protozoans, and mammalian tissue (25). Here we describe several FTA-based protocols for the extraction of fungal genomic DNA from yeasts and molds (25, 26), and a hitherto unpublished method for the ultra-rapid preparation of PCR-grade yeast genomic DNA. The two methods are summarized in Fig. 1.

2. Materials

1. Sabouraud's (SAC) medium (glucose-peptone agar containing 0.05 mg/ml chloramphenicol; Oxoid, Basingstoke, UK; catalog number PO0161). Store at 4°C.
2. Acid washed glass beads. Store at room temperature.
3. Sterile 1.5 ml type D screw-top microtubes.
4. Fast-Prep 24 instrument (MP Biomedicals Inc., Solon OH, USA).
5. Injection-quality sterile water, e.g., 10 ml snap-top ampoules. Store at room temperature, aliquot when opened and store aliquots at –20°C.
6. Whatman FTA filters, 4-spot Indicator classic cards. Store dry, at room temperature.
7. Harris sterile uni-core punches, 2.0 mm diameter (Whatman International Ltd., Banbury, UK) (see Note 1).
8. Self-healing Harris cutting mat (Fisher Scientific/Finnzymes) (see Note 2).
9. Sterile 0.2 ml thin-walled PCR tubes with flat caps (Bio-rad, Hercules, CA, USA).
10. TE wash reagent (10 mM Tris–HCl pH 7.5, 0.1 mM EDTA in sterile, distilled water). Store in 10 ml aliquots at room temperature.
11. Whatman FTA purification reagent, supplied in 500 ml volumes (Whatman International Ltd., Banbury, UK). Aliquot into 20 ml volumes in sterile tubes and store at room temperature.
12. Microwave.

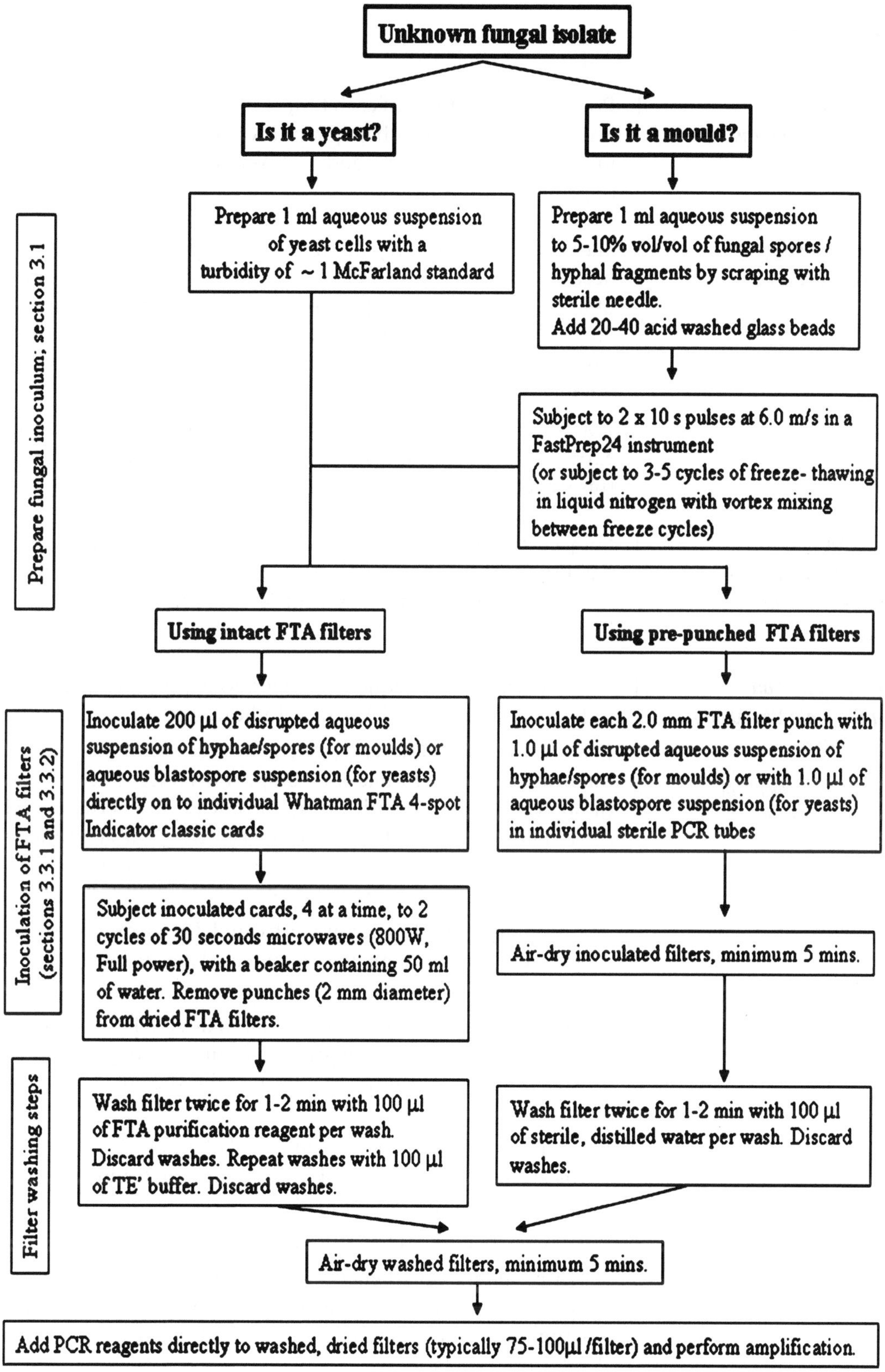

Fig. 1. Algorithm for the use of FTA filter papers for the extraction of genomic DNA from pure cultures of yeasts and molds.

3. Methods

3.1. Preparation of Fungal Cultures

1. Prepare pure cultures of each fungal isolate by inoculating onto SAC medium and incubate plates inverted at 30°C for a minimum of 48 h (for yeasts) and 1–2 weeks (depending on speed of growth for molds).

3.2. Harvesting of Fungal Cultures

1. Harvest cultures by scraping the surfaces of colonies with sterile needles/loops and suspending harvested fungi in 1 ml volumes of sterile water in screw-top microtubes. For yeast isolates, employ 25 μl of a heavy blastoconidial suspension; for molds, add a spore suspension and/or hyphal fragments to ~10% vol/vol for hyphal fragments, and 2% vol/vol for spores (see Note 3).
2. For harvested molds, add 20–40 acid washed glass beads per extract. Harvested molds and yeasts can be stored at –20°C at this stage.

3.3. Physical Disruption of Harvested Mold Extracts

1. Subject mold extracts to 2 × 10 s pulses at 6.0 m/s in a FastPrep24 instrument. This step can be replaced by 3–5 cycles of freeze–thawing in liquid nitrogen with vortex mixing between freeze cycles.

3.4. Purification of Genomic DNA Using FTA Filters

3.4.1. Using Whole FTA Cards for Archiving Important DNA Preparations

1. Apply yeast or mold aqueous suspensions (200 μl) directly to individual Whatman FTA 4-spot Indicator classic cards.
2. Place Whatman FTA cards, open, in a conventional kitchen microwave (800 W) while still damp, and subject to two cycles of 30 s on full power, with a pause of at least 30 s between each cycle. Filters can be subjected to microwaves four at a time, in the presence of a pyrex beaker containing 50 ml of sterile water to dissipate excess heat (see Note 4).
3. At this stage, dried, microwave-treated filters can be stored indefinitely at room temperature in sealed plastic bags containing a sachet of dessicant.
4. Remove punches (2 mm diameter) from dried FTA filters using a Harris uni-core punch and the self-healing cutting mat. Place a single filter punch directly into a sterile 0.2 ml thin-walled PCR tube.
5. Wash filters twice for 1 min each with 100 μl of Whatman FTA purification reagent, followed by 2 washes for 1 min each with 100 μl of TE buffer (see Note 5).
6. Air-dry washed filters for 5–10 min at room temperature.
7. Air-dried filters can then be directly subjected to molecular approaches. Typically we add PCR reagents (75–100 μl final reaction volumes) directly to the FTA filters in sterile 0.2 ml PCR tubes (see Note 6).

3.4.2. Using Pre-punched FTA Cards for Ultra Inexpensive DNA Preparation

The method detailed in Subheading 3.3, step 1 has been successfully used for the extraction of PCR-grade fungal genomic DNA from over 2,000 isolates of molds and yeasts encompassing over 150 different species of fungi ((25) and unpublished data). However, this method does have several drawbacks, including the price of individual FTA filters and the danger of cross-contamination from filter to filter when filter punches are removed using a reusable micro-punch (see Notes 2 and 3). This second limitation precludes the use of FTA technology for molecular diagnostic approaches in mycology, where the aim is to detect the presence of a fungal infection in normally sterile clinical samples rather than identify a fungus growing in pure culture. The use of pre-punched FTA filters as described below circumvents both of these limitations:

1. Remove individual 2.0 mm filter punches from an intact (non-inoculated), indicating FTA Classic Card using a sterile (not used previously for inoculated cards) Harris uni-core punch and the self-healing cutting mat in a laminar flow cabinet.
2. Store the individual filter punches (in excess of 500 punches can be removed per FTA card) dry in separate sterile 0.2 ml thin-walled PCR tubes until required.
3. Inoculate the individual filter punches with pre-prepared fungal extracts. For yeasts, the inoculum corresponds to 1 μl of an aqueous suspension of yeast blastospores with a turbidity of approximately 1 McFarland standard (see Note 7). For mold isolates, apply 1 μl of the physically disrupted extract preparation (from Subheading 3.2, step 1) directly to each 2.0 mm FTA filter punch.
4. For both yeasts and molds, air-dry inoculated filters for 5–10 min at room temperature.
5. Wash dried inoculated filters with two successive 2 min washes using 100 μl of sterile water for each wash (see Note 5).
6. Following washing, air-dry filters again at room temperature for 5–10 min.
7. Air-dried filters can then be directly subjected to molecular approaches. Typically we add PCR reagents (75–100 μl final reaction volumes) directly to the FTA filters in sterile 0.2 ml PCR tubes (Note 6).

4. Notes

1. When using the Harris uni-core micro-punch to remove filter punches from successive cards inoculated with different organisms, the punch must be cleaned between each filter

card. This can be achieved by taking consecutive (4, 5) punches from a non-inoculated FTA card or sterile paper towels that have been dampened with sterile and distilled water. Alternatively, the uni-core micro-punch tip can be treated by brief soaking in a proprietary product designed to destroy DNA and RNA.

2. When using the self-healing cutting mat to prepare filter punches from successive cards inoculated with different organisms, the cutting mat must be cleaned between each filter card. This can be achieved by wiping with sterile paper towels that have been dampened with sterile and distilled water or treating with a proprietary product designed to destroy DNA and RNA.

3. Any suspected ACDP (Advisory Committee on Dangerous Pathogens) category 3 fungal pathogens must be handled and hyphal fragments/spores must be harvested in an appropriate BSL3 containment facility. Here, we must also stress the need for rigorous inactivation of any potential category (ACDP) 3 fungal pathogens by extended heat treatment (at least 30 min at 95°C or higher) and subsequent culture-based verification of complete inactivation before such organisms are applied to FTA cards.

4. Our initial experiments with FTA technology demonstrated that FTA filter matrix mediated inactivation of fungi was largely dependent on the dose of organism applied, presumably due to the large size of fungal cells, and the possibility that not all cells were in direct contact with the FTA paper when concentrated inocula were employed. Thus, we employ microwave treatment as a useful modification of current FTA methodologies since it allows total organism inactivation without the need for laborious titration of applied suspensions and also effectively dries inoculated cards (25).

5. Tests in our laboratory have not demonstrated a significant difference (as assessed by PCR success rates, and amplicon yields; data not shown) between filters washed with FTA wash reagents and those washed only with sterile water (see, for example, Fig. 2A in ref. (26)). Thus our current protocols employ two successive 2 min washes using 100 μl of sterile water for each wash.

6. We have now employed this method for over 500 isolates (over 100 species) of filamentous fungi with excellent success rates (>95% of isolates yielded PCR-grade genomic DNA as assessed by amplification success using either PCR primers targeting multicopy genes in the nuclear ribosomal RNA gene cassette and also single copy genes; ref. 26 and unpublished data). Although genomic DNA prepared in this way cannot be archived (only a single filter punch is inoculated as opposed to an entire FTA card), this method is over 500 times less expen-

sive than when intact FTA cards and FTA purification reagents are used (26) and is ideal for the rapid molecular analyses of large numbers of clinical isolates of molds and yeasts.

7. Tests in our laboratory have shown that yeast genomic DNA can be amplified by PCR without the use of FTA filters. Experiments showed that intact yeast cells could be added directly to PCRs at a wide range of cell concentrations without inhibitory effects on the PCR. Our current protocols for PCR amplification of yeast genomic DNA involve the preparation of an aqueous suspension of yeast cells corresponding to 0.5 MacFarland standard and programming of 100 μl PCRs with 1 μl of this suspension. Our PCRs use a hot-start Taq polymerase, requiring an initial activation step at 95°C for 15 min, which is sufficient to lyse the yeast cells and release genomic DNA (unpublished data).

Acknowledgments

We thank the other staff of the UK National Mycology Reference Laboratory for their interest in these approaches and for providing much of the data on which these approaches are based. We are also indebted to Whatman International for supplying the FTA cards and wash reagents used for these studies and to Haroun Shah and Dunstan Rajendram for their invaluable advice on FTA applications.

References

1. Nucci M, Marr KA (2005) Emerging fungal diseases. Clin Infect Dis 41:521–526
2. Richardson M, Lass-Flörl C (2008) Changing epidemiology of systemic fungal infections. Clin Microbiol Infect 14(Suppl 4):5–24
3. Linton CJ, Borman AM, Cheung G, Holmes AD, Szekely A, Palmer MD, Bridge PD, Campbell CK, Johnson EM (2007) Molecular identification of unusual pathogenic yeast isolates by Large Ribosomal Subunit gene sequencing: 2 years experience at the UK Mycology Reference Laboratory. J Clin Microbiol 45:1152–1158
4. Borman AM, Linton CJ, Miles S-J, Johnson EM (2008) Molecular identification of pathogenic fungi. J Antimicrob Chemother 61:i7–i12
5. Tavanti A, Davidson AD, Gow NAR, Maiden MCJ, Odds FC (2005) *Candida orthopsilosis* and *Candida metapsilosis* spp. nov. To replace *Candida parapsilosis* groups II and III. J Clin Microbiol 43:284–292
6. Gilgado F, Cano J, Gené J, Guarro J (2005) Molecular phylogeny of the *Pseudallescheria boydii* species complex: proposal of two new species. J Clin Microbiol 43:4930–4942
7. O'Donnell K, Sutton DA, Fothergill A, McCarthy D, Rinaldi MG, Brandt ME, Zhang N, Geiser DM (2008) Molecular phylogenetic diversity, multilocus haplotype nomenclature, and in vitro antifungal resistance within the *Fusarium solani* species complex. J Clin Microbiol 46:2477–2490
8. Balajee SA, Kano R, Baddley JW, Moser SA, Marr KA, Alexander BD, Andes D, Kontoyiannis DP, Perrone G, Peterson S, Brandt ME, Pappas PG, Chiller T (2009) Molecular identification of Aspergillus species collected from the transplant-associated surveillance network. J Clin Microbiol 47:3138–3141
9. Balajee SA, Weaver M, Imhof A, Gribskov J, Marr KA (2004) *Aspergillus fumigatus* variant with decreased susceptibility to multiple antifungals. Antimicrob Agents Chemother 48:1197–1203

10. Balajee SA, Gribskov J, Brandt M, Ito J, Fothergill A, Marr KA (2005) Mistaken identity: *Neosartorya pseudofischeri* and its anamorph masquerading as *Aspergillus fumigatus*. J Clin Microbiol 43:5996–5999
11. Alcazar-Fuoli L, Mellado E, Alastruey-Izquierdo A, Cuenca-Estrella M, Rodriguez-Tudela JL (2008) *Aspergillus* section *Fumigati*: antifungal susceptibility patterns and sequence-based identification. Antimicrob Agents Chemother 52:1244–1251
12. Borman AM, Petch R, Linton CJ, Palmer MD, Bridge PD, Johnson EM (2008) Candida nivariensis, an emerging pathogenic yeast with multi-drug resistance to antifungal agents. J Clin Microbiol 46:933–938
13. Houbraken J, Verweij PE, Rijs AJMM, Borman AM, Samson RA (2010) Identification of *Paecilomyces variotii* in clinical samples and settings. J Clin Microbiol 48:2754–2761
14. Tang C, Holden D, Aufauvre-Brown A, Cohen J (1993) The detections of *Aspergillus* spp. by the polymerase chain reaction and its evaluation of bronchealveolar lavage fluid. Am Rev Respir Dis 148:1313–1317
15. Al-Samarrai TH, Schmid J (2000) A simple method for extraction of fungal genomic *DNA*. Lett Appl Microbiol 30:53–56
16. Cenis JL (1992) Rapid extraction of fungal DNA for PCR amplification. Nucleic Acids Res 20:2380
17. Dean TR, Betancourt D, Menetrez MY (2004) A rapid DNA extraction method for PCR identification of fungal indoor air contaminants. J Microbiol Methods 56:431–434
18. Faggi E, Pini G, Campisi E (2005) Use of magnetic beads to extract fungal DNA. Mycoses 48:3–7
19. Guo JR, Schneider F, Abd-Elsalam KA, Vereet JA (2005) Rapid and efficient extraction of genomic DNA from different phytopathogenic fungi using DNAzol reagent. Biotechnol Lett 27:3–6
20. Hopfer RL, Walden P, Setterquist S, Highsmith WE (1993) Detection and differentiation of fungi in clinical specimens using polymerase chain recation amplification and restriction enzyme analysis. J Med Vet Mycol 31:65–75
21. Jin J, Lee Y-K, Wickes BL (2004) Simple chemical extraction method for DNA isolation from *Aspergillus fumigatus* and other *Aspergillus* species. J Clin Microbiol 42:4293–4296
22. Muller F-MC, Werner KE, Kasai M, Francesconi A, Chanock SJ, Walsh TJ (1998) Rapid extraction of genomic DNA from medically important yeasts and filamentous fungi by high-speed cell disruption. J Clin Microbiol 36:1625–1629
23. Van Burik JA, Schreckhise RW, White TC, Bowden RA, Myerson D (1998) Comparison of six extraction techniques for isolation of DNA from filamentous fungi. Med Mycol 36:299–303
24. Liu D, Coloe S, Baird R, Pederson J (2000) Application of PCR to the identification of dermatophyte fungi. J Med Microbiol 49:493–497
25. Borman AM, Linton CJ, Miles S-J, Campbell CK, Johnson EM (2006) Ultra-rapid preparation of total genomic DNA from isolates of yeast and mould using Whatman FTA filter paper technology—a re-usable DNA archiving system. Med Mycol 44:389–398
26. Borman AM, Fraser M, Linton CJ, Palmer MD, Johnson EM (2010) An improved method for the preparation of total genomic DNA from isolates of yeast and mould using Whatman FTA filter papers. Mycopathologia 169:445–449

Chapter 4

APEX DNA Microarray for the Identification of Pathogenic Fungi

Arianna Tavanti, Stefano Landi, and Sonia Senesi

Abstract

The identification of fungal pathogens, though continuously improving, is still time-consuming and often inadequate for ensuring an early targeted therapy, which may be crucial for the treatment of invasive mycoses. Here, we describe a DNA-microarray system based on the arrayed-primer extension (APEX) technique for a rapid identification of pathogenic fungi, which represents a critical step in medical practice.

Key words: DNA microarray, APEX, Molecular identification, Pathogenic fungi, ITS region, PCR

1. Introduction

Rapid diagnosis of invasive mycoses is an essential step for the prompt management of infection with tailored antifungal treatments (1). Conventional laboratory methods for the identification of fungal pathogens are still time-consuming and often inadequate for ensuring early targeted therapy, especially for uncommon or newly identified fungal species. This chapter is focused on describing the development and utilization of an oligonucleotide array for the unambiguous identification of 24 fungi, belonging to ten diverse genera, including closely related species still undistinguishable by conventional methods (2). The oligonucleotide probes used in this microarray are complementary to the sequence variation in the internal transcribed spacer 1 and 2 (ITS1 and ITS2) region of each species (2).

Louise O'Connor and Barry Glynn (eds.), *Fungal Diagnostics: Methods and Protocols*, Methods in Molecular Biology, vol. 968, DOI 10.1007/978-1-62703-257-5_4, © Springer Science+Business Media New York 2013

A single PCR is used to amplify ITS region in all fungi, thanks to the use of universal primers ITS1 and ITS4 (3), which are positioned externally to the ITS region, on highly conserved rRNA gene regions. The ITS region is perhaps the most widely analyzed DNA sequence in fungi and it has typically been used for molecular species-level systematics. By the arrayed-primer extension (APEX) technique, generation of signals for fungal species identification does not require direct labeling of PCR products. Indeed, this method relies both on specific hybridization between probes and their targets, like conventional microarray-based methods, and on the specific extension of a single base at the 3′ side of the anchored probe (4–6). In other words, APEX provides a further level of analysis and thus, at least theoretically, should give more clear-cut results and reduce misclassification of fungal species than those of hybridization-only array-based methods. Optimization of the detection system represents a key step for ensuring the reproducibility, specificity, and sensitivity of the microarray results.

2. Materials

2.1. Genomic DNA Extraction from Yeast and Molds

Sabouraud dextrose broth and agar, potato dextrose agar (molds). Tween 20, sterile saline; glass beads (0.45–0.52 mm diameter); lysis buffer (100 mM Tris–HCl, pH 8.0, 2% (vol/vol) Triton X-100, 1%(wt/vol) sodium dodecyl sulfate, 1 mM EDTA) 1:1 (vol/vol) phenol–chloroform; TE (10 mM Tris–HCl, pH 8.0, 1 mM EDTA); ethanol; RNAse; proteinase K (20 mg/ml stock solution) isopropanol; and 4 M ammonium acetate.

2.2. ITS Amplification

ITS1 (5′-TCCGTAGGTGAACCTGCGG-3′) and ITS4 (5′-TCCT CCGCTTATTGATATGC-3′) primers (5 μM); deoxynucleoside triphosphate mix (2 mM dATP, 2 mM dCTP, 2 mM dGTP, 1.7 mM dTTP, and 0.03 mM dUTP), 10× magnesium-free buffer, 25 mM $MgCl_2$, and 5 U/μl Hot Fire DNA polymerase. PX2 thermal cycler.

2.3. Microarray Setup

5′ C-6 Aminolinker-modified oligonucleotides (C-6 oligonucleotides) were designed by matching a region of about 35 bp within the *ITS1* amplification products to the sequences of ITS (2). Each C-6 oligonucleotide was designed in order to exclusively add a uracyl (fluorescein-ddUTP) at its 3′ end during the single-base extension reaction. All C-6 oligonucleotides were spotted onto silanized slides by Asperbio (Tartu, Estonia) (see Note 1).

For each fungal species, sets of oligonucleotides were selected among those previously designed on the basis of their discriminatory power and lack of cross-reactivity (2).

2.4. Purification, Hybridization of PCR Products on the Chip, and Signal Detection

The following reagents were used: Millipore Microcon Y30 columns for the purification of PCR products; uracil N-glycosylase (UNG) for fragmentation; shrimp alkaline phosphatase for dNTPs removal. For single base primer extension a mixture of fluorescein-ddUTP, cyanine 3-ddATP, ROX-ddCTP, cyanine 5-ddGTP (50 pmol each), and 5 U of Thermo Sequenase (diluted in its dilution buffer) and Thermo Sequenase 10× Reaction Buffer were used. The incubation was carried out in a final volume of 20 μL. For imaging capture we used SlowFade® Light Antifade Reagent (Molecular Probes, Eugene, OR, USA) and Genorama™-003 four-colors detector equipped with Genorama image analysis software (Asper Biotech, Tartu, Estonia) (see Note 2).

3. Methods

3.1. Genomic DNA Extraction from Yeast and Molds

1. Grow fungi in Sabouraud dextrose broth (yeast) or on potato dextrose agar (molds) for 24 h (yeast) or until colonies are fully grown on plate at 30°C.
2. Harvest yeast in stationary phase and prepare mold suspensions by collecting conidia from 72 h potato dextrose agar plate colonies washed with 5 ml of 0.1% Tween 20–sterile saline (see Note 3).
3. Lyse both yeast and conidia by vortexing the pellet for 3 min with 0.3 g glass beads (0.45–0.52-mm diameter) in 200 μl of lysis buffer (100 mM Tris–HCl, pH 8.0, 2% (vol/vol) Triton X-100, 1% (wt/vol) sodium dodecyl sulfate, and 1 mM EDTA) and 200 μl of 1:1 (vol/vol) phenol–chloroform (see Note 4).
4. After the pellet has been fully resuspended, add 200 μl of TE (10 mM Tris–HCl, pH 8.0, 1 mM EDTA) to the lysate, mix gently, and centrifuge the mixture at full speed for 10 min.
5. Transfer the aqueous phase to a new tube. Precipitate DNA by adding 1 ml of ethanol (see Note 5). Mix gently by inversion.
6. Centrifuge samples at full speed for 10 min and suspend the pellet in 400 μl of TE containing 100 μg RNase. Incubate the mixture for 1 h at 37°C.
7. Add proteinase K (5 μl of a 20-mg/ml stock solution) and incubate for 1 h at 65°C and then for 30 min at 72°C.
8. Repeat treatment with phenol–chloroform and precipitate DNA with 1 ml of isopropanol and 10 μl of 4 M ammonium acetate.
9. Mix gently by inversion, centrifuge samples at full speed for 10 min, remove supernatant, dry the pellet, and dissolve it in 50 μl of TE, pH 8.0.

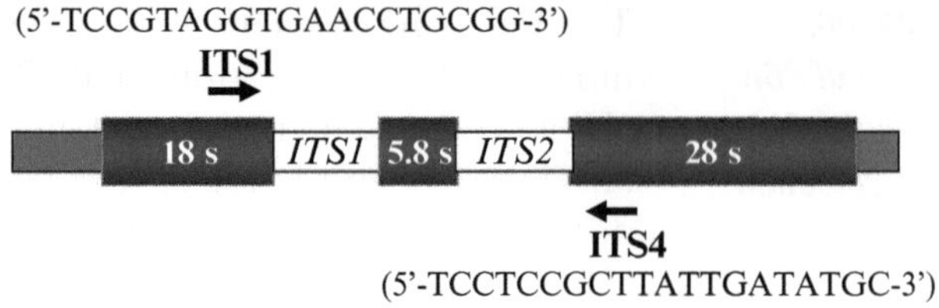

Fig. 1. Schematic representation of the ITS region and universal primers (3) used for amplification.

3.2. ITS Amplification

1. Use the universal fungal primers ITS1 and ITS4 (3) to amplify the non-coding ITS regions (*ITS1* and *ITS2*) as well as the 5.8S rRNA gene, positioned between the ITS regions (Fig. 1).
2. Set up a reaction mixture (20 μl) containing fungal DNA (100 ng), 0.4 μl primers (5 μM), 2 μl of deoxynucleoside triphosphate mix (2 mM dATP, 2 mM dCTP, 2 mM dGTP, 1.7 mM dTTP, and 0.03 mM dUTP), 2 μl magnesium-free 10X buffer, 2 μl $MgCl_2$ (25 mM), and 0.2 μl hot start Taq DNA polymerase (5 U/μl).
3. Include a negative control in all PCR experiments, which consists of an equal volume of water replacing the DNA template.
4. PCR amplification was performed using the following conditions: 94°C for 15 min (hot start); 35 cycles of denaturation at 95°C (1 min), annealing at 56°C (30 s), and elongation at 72°C (75 s); and a final extension step at 72°C (10 min).
5. Separate the amplified DNA products by electrophoresis in a 1% agarose gel containing ethidium bromide (0.5 mg/ml) and TAE (40 mM Tris acetate (pH 8.0), 1 mM EDTA) as the running buffer.
6. Use a 100 bp DNA ladder as a molecular size marker.
7. Visualize DNA bands by UV transillumination (Figs. 2 and 3).

3.3. Purification, Amplicon Hybridization, and Signal Detection

1. Purify PCR products and concentrate PCR products using spin columns (Millipore Microcon Y30 or equivalent).
2. Reduce amplicon sizes by fragmentation to improve hybridization with the arrayed oligonucleotides. This is achieved by treating the column eluate samples (15 μl) with 1 U UNG and 1 U shrimp alkaline phosphatase. Incubate the mixtures at 37°C for 1.5 h and at 95°C for 30 min. At this temperature, DNA with abasic sites is denatured and fragmented, whereas UNG and shrimp alkaline phosphatase are inactivated. Add aliquots of the treated amplicons (9 μl) to a reaction mixture containing fluorescein-ddUTP, cyanine 3-ddATP, ROX-ddCTP, cyanine 5-ddGTP (4 × 50 pmol), 10× buffer, and 4 U

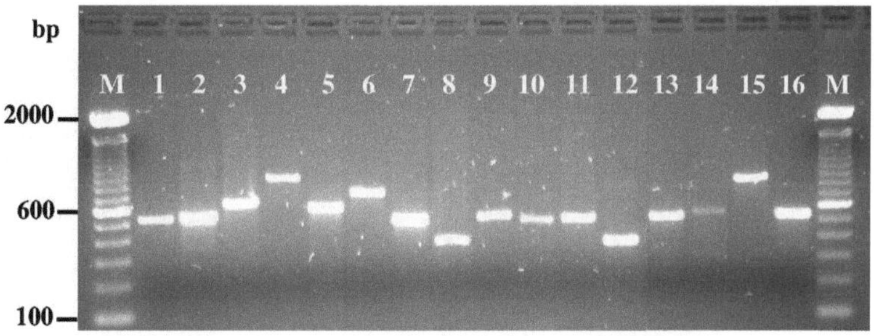

Fig. 2. ITS PCR products of 16 yeast species. Lane 1, *Candida albicans*; Lane 2, *C. dubliniensis*; Lane 3, *C. famata*; Lane 4, *C. glabrata*; Lane 5, *C. guilliermondii*, Lane 6, *C. kefyr*, lane 7, *C. krusei*; Lane 8, *C. lusitaniae*; Lane 9, *C. metapsilosis*; Lane 10, *C. orthopsilosis*; Lane 11; *C. parapsilosis*; Lane 12, *C. pulcherrima*; Lane 13, *C. tropicalis*; Lane 14, *Cryptococcus neoformans*; Lane 15, *Saccharomyces cerevisiae*; Lane 16, *Trichosporon cutaneum*. Lane M, 100 bp DNA ladder.

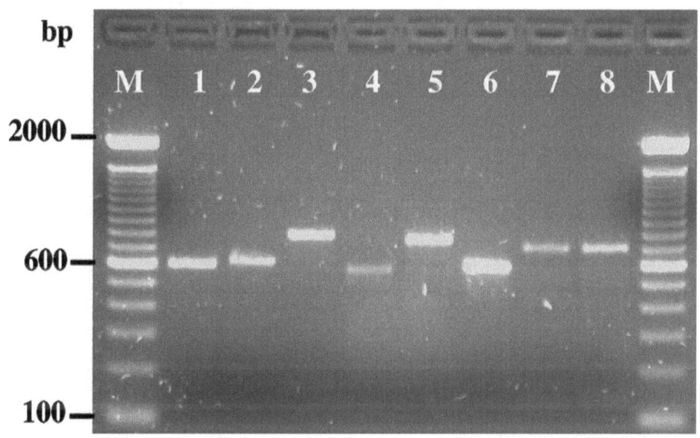

Fig. 3. ITS PCR products of 8 mold species. Lane 1, *Aspergillus fumigatus*; Lane 2, *A. terreus;* Lane 3, *Epidermophyton floccosum*; Lane 4, *Fusarium solani*; Lane 5, *Microsporum canis*, Lane 6, *Penicillium marneffei*; Lane 7, *Trichophyton rubrum*; Lane 8, *Trichophyton tonsurnas*. Lane M, 100 bp DNA ladder.

of Thermo Sequenase diluted in the provided dilution buffer to give a final volume of 20 μl.

3. Pre-wash the microarray slides with 100 mM NaOH and rinse with distilled water at 95°C.
4. Quickly place mixtures onto the spotted slide, cover with a coverslip, and incubate at 58°C for 10 min.
5. Wash the slides with distilled water at 95°C to remove both non-hybridized PCR products and unincorporated labeled dideoxynucleoside triphosphates. The rinse should last only the time the coverslip is placed.

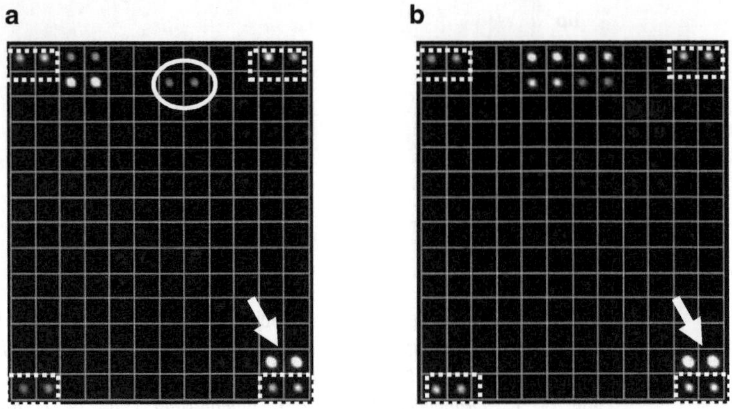

c

Contr ●●	*C. albicans 1* ●●	*C. dubliniensis 1* ●●	*C. dubliniensis 3* ●●	*C. tropicalis1* ●●	Contr ●●
A channel ●●	*C. albicans 2* ●●	*C. dubliniensis 2* ●●	*C. dubliniensis 4* ●●	*C. tropicalis2* ●●	C channel ●●
	C. parapsilosis 1 ●●	*C. metapsilosis 1* ●●	*C. orthopsilosis 1* ●●		
	C. parapsilosis 2 ●●	*C. metapsilosis 2* ●●	*C. orthopsilosis 2* ●●		*C. krusei* ●●
	C. kefyr 2 ●●	*C. kefyr 4* ●●			*C. pulcherrima* ●●
C. kefyr1 ●●	*C. kefyr 3* ●●		*C. lusitaniae1* ●●	*C. lusitaniae2* ●●	
C. famata 1 ●●	*C. guilliermondii 1* ●●				
C. famata 2 ●●	*C. guilliermondii 2* ●●		*C. neoformans1* ●●		*S. cerevisiae 1* ●●
	F. solani 1 ●●	*A. fumigatus 1* ●●	*A. terreus 1* ●●	*P. marneffei 1* ●●	*P. marneffei 3* ●●
	F. solani 2 ●●	*A. fumigatus 2* ●●	*A. terreus 2* ●●	*P. marneffei 2* ●●	*P. marneffei 4* ●●
M. canis 1 ●●	*M. canis 3* ●●	*T. rubrum 1* ●●	*T. rubrum 3* ●●		*T. tonsurans 1* ●●
M. canis 2 ●●	*M. canis 4* ●●	*T. rubrum 2* ●●	*T. rubrum 4* ●●	*T. rubrum 5* ●●	*T. tonsurans 2* ●●
G channel ●●	*E. floccosum 1* ●●	*E. floccosum 3* ●●			T channel ●●
Contr ●●	*E. floccosum2* ●●			*T. cutaneum* ●●	Contr ●●

Fig. 4. Representative Cy3 fluorescence image snapshots from Genorama Basecaller software showing the identification of *Candida albicans* (**a**) and *C. dubliniensis* (**b**) by the array. No signals were detected in the Cy5 channel (i.e., the non-U channel, which is not displayed for brevity). (**c**) Layout of chip capture probes (for probe sequences, see ref. (2)). Control probes are positioned at the corners of each panel, indicated by *dashed lines*. Positive control for the uridine channel is indicated by the *arrow*. When *C. albicans* is present is the sample, a cross reaction with one of the four probes of *C. dublininesis* is observed (*C. dubliniensis* 4, *white circle*). However, a clear identification of both species is obtained.

6. Add a drop of antifade reagent to prevent fluorescein signal fading.
7. Capture slide images with a four-color detector (Fig. 4).
8. Measure fluorescence intensities at each position and convert them to base calls, according to Genorama image analysis and genotyping software.

3.4. Interpretation of Results

Two main criteria are required for species identification:

1. The extension has to yield a signal on the U channel, as the two species-specific probes were designed to extend uridine only (see Notes 6 and 7).
2. Both APEX probes must give a signal, then 4 signals must be positive at the U-channel, since each probe is spotted in duplicated on the slide.

4. Notes

1. In our opinion, excellent results were obtained using silanized slides provided by Asperbio (Tartu, Estonia).
2. Purification and concentration of PCR products gave the best results using Millipore Microcon Y30 columns. In our experience best results were obtained using UNG from EPICENTRE, Madison, WI, USA; Thermosequenase from USB Amersham plc, Buckinghamshire, UK; and shrimp alkaline phosphatise (sAP) from Amersham Biosciences, Milwaukee, WI, USA. In addition, SlowFade® Light Antifade Reagent (Molecular Probes, Eugene, OR, USA) and Genorama™-003 four-color detector equipped with Genorama image analysis software (Asper Biotech, Tartu, Estonia) provided excellent results.
3. In case of filamentous fungi, conidia are more easily collected than fungal mycelium, which is often invading the agar plate, while a conidial suspension can be promptly prepared using a Tween 20 solution.
4. Check that no beads remain along the eppendorf tip border, since this may prevent a proper closure of the microcentrifuge tube cap, leading to potential leaks. Covering the eppendorf tip with parafilm usually helps preventing phenol leaking.
5. At this point a transparent, flocculating structure could be visualized within the microcentrifuge tube.
6. Although A, G, and C signals are not expected to be incorporated in the APEX reaction, it is important to include ddA, ddC, and ddG in the reaction mixture, since this will provide a further specificity control to detect any unspecific extensions.
7. To ensure quality control, it is recommended to spot each oligonucleotide in duplicate and to include internal positive controls (self-extending oligonculeotides already spotted at the corners of the array) to verify that the intensities of the four channels (A, C, U, and G) are equilibrated.

Acknowledgments

This work was supported by research grant 200985J87J_003 from the "Ministero dell'Istruzione, dell'Università e della Ricerca" (MIUR) and by an "Associazione Italiana Ricerca sul Cancro" (AIRC) regional grant.

References

1. Landlinger C, Preuner S, Willinger B et al (2009) Species-specific identification of a wide range of clinically relevant fungal pathogens by use of Luminex xMAP technology. J Clin Microbiol 47:1063–1073
2. Campa D, Tavanti A, Gemignani F et al (2008) DNA microarray based on arrayed-primer extension technique for identification of pathogenic fungi responsible for invasive and superficial mycoses. J Clin Microbiol 46:909–915
3. White T, Bruns T, Lee S et al (1990) Amplification and direct sequencing of fungal ribosomal RNA genes for phylogenetics. In: Innis M, Gelfand D, Sninsky J, White T (eds) PCR protocols: a guide to methods and application. Academic Press, San Diego, CA, pp 315–322
4. Gemignani F, Landi S, Chabrier A et al (2004) Generation of a DNA microarray for determination of E6 natural variants of human papillomavirus type 16. J Virol Methods 119:95–102
5. Gemignani F, Perra C, Landi S et al (2002) Reliable detection of beta-thalassemia and G6PD mutations by a DNA microarray. Clin Chem 48:2051–2054
6. Landi S, Gemignani F, Gioia-Patricola L et al (2003) Evaluation of a microarray for genotyping polymorphisms related to xenobiotic metabolism and DNA repair. Biotechniques 35:816–827

Chapter 5

Microscopic Detection of Yeasts Using Fluorescence In Situ Hybridization

João Inácio and Maria da Luz Martins

Abstract

Fluorescence in situ hybridization (FISH) has been widely used for the detection and identification of microorganisms in their natural environments. In this chapter we describe the use of a simple FISH-based protocol to detect and identify clinically relevant yeast species in culture and biological samples using *Cryptococcus neoformans* as a model. After fixation of cells with paraformaldehyde, the same are embedded in hybridization buffer containing specific fluorochrome-labeled oligonucleotide probes. After incubation and a subsequent washing step for removing unbound probes, samples are analyzed by epifluorescence microscopy.

Key words: *Cryptococcus neoformans*, Fluorescence in situ hybridization, Ribosomal RNA, Epifluorescence microscopy, Cy3-labeled DNA probes, Yeast detection and identification

1. Introduction

Fluorescence in situ hybridization (FISH) has been widely used for the detection and identification of microorganisms with emphasis on prokaryotic organisms, allowing the direct microscopic visualization and quantification of individual cells in their natural environments including biological samples such as host tissues (1–3). FISH has also been used to detect and identify fungal organisms including clinically relevant yeasts such as *Candida* and *Cryptococcus* species (4, 5), food and crop-associated yeasts (6–8), and environmental fungi particularly of phylloplane inhabitants (9–11). Briefly, conducting FISH assays first involves the fixation of cells, or samples, using polymerizing or precipitating agents such as paraformaldehyde or alcoholic solutions, respectively. Eventually cells must

Louise O'Connor and Barry Glynn (eds.), *Fungal Diagnostics: Methods and Protocols*, Methods in Molecular Biology, vol. 968, DOI 10.1007/978-1-62703-257-5_5,

undergo an additional process of enzymatic permeabilization to facilitate the access of the probes to intracellular target sites. Ribosomal RNA is the most commonly used gene target for designing FISH probes since it is present in all cellular organisms, is genetically stable, is a multicopy target, and contains both conserved and variable regions allowing the design of probes targeting a wide range of individual species, genera, or other taxonomic units (1, 12, 13). In addition, another important factor is the large number of rRNA nucleotide sequences available from public databases. After mixture and incubation with fluorochrome-labeled oligonucleotide, polynucleotide or peptide nucleic acid (PNA) probes, and a subsequent washing step for removing unbound probes, samples can then be analyzed by epifluorescence or laser scanning microscopy or flow cytometry (1). In this chapter we describe the use of a simple FISH-based protocol to detect and identify clinically relevant yeast species in culture and biological samples, using *Cryptococcus neoformans* as a model. This yeast is present in the environment worldwide and can cause disseminated disease, particularly in patients with depressed cell-mediated immunity, the meninges being the most common site of extrapulmonary infection (14). The conventional diagnosis of cryptococcosis is based on the direct microscopic examination of clinical samples such as cerebrospinal fluid (CSF) and on the culture and subsequent phenotypic identification of the etiological agent.

FISH assays are described below for detecting and identifying *Cryptococcus neoformans* cells in both pure culture and in infected CSF samples.

2. Materials

1. Prepare all solutions using Milli-Q or deionized water.
2. Type strain of *Cryptococcus neoformans* obtained from the Portuguese Yeast Culture Collection, Caparica, Portugal, with the number PYCC 3957^{T} (see Note 1).
3. CSF samples from HIV-positive patients with clinical diagnosis of cryptococcosis (5) (see Note 2).
4. Yeast-Malt (YM) broth: 0.3% w/v malt extract, 0.2% yeast extract, 0.5% peptone, 1% glucose. Weigh all components to a glass beaker, dissolve in the specified amount of water, and distribute into 5 mL test tubes with metal or plastic caps. Sterilize in the autoclave (15 min, 121°C) and store at 4°C until use.
5. Probe specific for *Cryptococcus neoformans*: Cne205 (5′ GTC GCG TTA CTT GGG AGT-3′), targeting the 26S rRNA and labeled at the 5′ end with the fluorochrome Cy3. This probe also targets the closely related *Cryptococcus gattii* (5) (see Note 3).

6. Probes EUK516 (5′-ACC AGA CTT GCC CTC C-3′; targeting eukaryotic organisms) and EUB338 (5′-CTG CTG CCT CCC GTA GGA GT-3′; targeting bacteria), labeled at their 5′ end with the fluorochrome Cy3, are used as positive and negative hybridization controls, respectively (15).
7. Probe stocks are frequently delivered, lyophilized, and need to be diluted with sterile water according to the manufacturer's instructions. Stock solutions can be prepared at a standard concentration of 100 pmol/μL and stored in the dark at −20°C. Aliquots of each Cy3-labeled probe working solutions are prepared from stock solutions at a concentration of 30 ng/μL (approximately 4–5 pmol/μL for an 20-mer probe) in water and stored also in the dark at −20°C (see Note 4).
8. Phosphate-buffered saline: 137 mM NaCl, 2.7 mM KCl, 8 mM Na_2HPO_4, 1.5 mM KH_2PO_4, pH 7.2. Weigh all components into a glass beaker and dissolve in 800 mL water. Adjust the pH to 7.2 with HCl and/or NaOH solutions. Add distilled water to a total volume of 1 L, dispense the solution into aliquots, and sterilize by autoclaving (20 min, 121°C). Store at room temperature.
9. To prepare a 4% solution of paraformaldehyde (4% PFA) weigh 4 g of paraformaldehyde powder into 90 mL PBS 1× and heat in a water bath to 60°C for 1–2 h until suspension is clear (the addition of some drops of 10 N NaOH helps the solubilization). After clearing and cool down, adjust to pH 7.0 and bring the volume to 100 mL with PBS. Sterilize by passing through a syringe-adapted 0.2 μm filter (see Note 5).
10. To prepare a stock solution of 5 M NaCl, for hybridization, weigh 29.22 g NaCl to a glass beaker and bring to a final volume of 100 mL water. Sterilize by autoclaving (15 min, 121°C) and store at room temperature.
11. A stock solution of 10% sodium dodecyl sulfate (SDS) is required for hybridization. Weigh 10 g SDS to a glass beaker and bring to a final volume of 100 mL water (see Note 6). Store at room temperature.
12. Prepare a stock solution of 1 M Tris–HCl by weigh 12.1 g Tris-base (MW 121.14) to a glass beaker and dissolve in 90 mL water. Adjust to pH 7.2 with concentrated HCl, bring the volume to 100 mL water, sterilize by autoclaving (15 min, 121°C), and store at room temperature.
13. Hybridization buffer should be prepared as follows: 20 mM Tris–HCl, pH 7.2, 0.9 M NaCl, 0.01% w/v SDS, and an amount of formamide previously optimized for each probe (10% for Cne205 and 20% for EUK516 and EUB338) (see Note 7). For example, for preparing 2 mL of hybridization buffer for use in FISH assays with Cne205 probe, mix 360 μL 5 M NaCl, 40 μL 1 M Tris–HCl, 2 μL 10% SDS, 200 μL formamide, and 1,398 μL

Table 1
Equivalence between the concentration of formamide in hybridization buffer and the correspondent concentrations of sodium chloride and EDTA in washing buffer

Formamide in hybridization buffer (%)	NaCl in washing buffer (M)	Volume of NaCl and EDTA stock solutions added to the washing buffer (for a final volume of 50 mL)	
		5 M NaCl (μL)	0.5 M EDTA, pH 8 (μL)
0	0.900	9,000	0
5	0.636	6,300	0
10	0.450	4,500	0
15	0.318	3,180	0
20	0.225	2,150	500
25	0.159	1,490	500
30	0.112	1,020	500
35	0.080	700	500
40	0.056	460	500
45	0.040	300	500
50	0.028	180	500

water in a 2 mL microtube. The proportions of formamide and water can be adjusted in order to prepare buffers with other formamide concentrations (see Note 8).

14. Prepare a stock solution of 0.5 M EDTA by weighing 186.1 g $Na_2EDTA·2H_2O$ to a glass beaker and dissolving in 800 mL water using stirring. Add 20 g of NaOH pellets and adjust to pH 8 (see Note 9). Bring the volume to 1,000 mL with water, sterilize by autoclaving (15 min, 121°C), and store at room temperature.

15. Washing buffer should be prepared as follows: 20 mM Tris–HCl, pH 7.2, 0.01% w/v SDS, and an amount of EDTA and NaCl according to Table 1, depending on the optimal formamide concentration used in the respective hybridization buffer. For example, for preparing 50 mL of washing buffer for use in FISH assays with EUK516 and EUB338 probes (which use 20% formamide in hybridization buffer), mix 1 mL 1 M Tris–HCl, 50 μL 10% SDS, 2,150 μL 5 M NaCl, and 500 μL 0.5 M EDTA solutions and adjust to the final volume of 50 mL with water (Table 1) (see Note 10).

16. Fluorescence microscope equipped with a 100× oil objective, a 100 W mercury lamp, and an appropriate filter set for Cy3 fluorochrome (see Note 11).
17. Immersion oil, which must comply with the microscope 100× objective, and be nonfluorescent.
18. Anti-fading solution such as Vectashield solution (Vector, Burlingame, Calif.).

3. Methods

3.1. Fixation of Cultures and CSF Samples

1. Grow the pure culture of *Cryptococcus neoformans* PYCC 3957^T aerobically under continuous shaking in YM broth at 25°C.
2. Harvest cells in the early exponential growth phase (optical density of about 1.0 at 600 nm) by centrifugation of 500 μL culture aliquots for 5 min at 15,000 × *g* (see Note 12).
3. Wash cell pellets with 500 μL PBS and centrifuge for 5 min at 15,000 × *g*.
4. Resuspend the cells in 500 μL 4% PFA and incubate for 4 h at 4°C (see Note 13).
5. After fixation, recover cells by centrifugation (5 min at 15,000 × *g*), wash twice with 500 μL PBS, and resuspend in 50 μL of PBS and 50 μL of cold absolute ethanol (see Note 14). Store at −20°C until use.
6. For CSF samples, centrifuge 1 mL sample at 15,000 × *g* for 5 min to pellet yeast cells, wash once with 500 μL of PBS, fix with 4% PFA, and store in PBS/ethanol as described above (see Note 15).

3.2. Hybridization

1. Add 2 μL of the fixed *Cryptococcus neoformans* suspensions, originally collected from a pure culture, into the bottom of three 1.5 mL microtubes (see Note 16).
2. Add 20 μL of hybridization buffer (containing 10% formamide when testing Cne205 and 20% when testing EUK516 and EUB338 probes) to each microtube.
3. Add 1 μL of each Cy3-labeled DNA probe (Cne205, EUK516 and EUB338) to the correspondent microtube, gently mix, and incubate in the dark at 46°C for 2 h (with the lids closed) (see Note 17).
4. The fixed *Cryptococcus neoformans* suspensions collected from CSF samples are centrifuged (15,000 × *g* for 5 min) and the entire pellets are resuspended in 20 μL of hybridization buffer (containing 10% formamide). The samples are tested only with Cne205 probe as above (see Note 18).

3.3. Stringent Washing and Visualization of Results

1. After incubation, cells are pelleted by centrifugation (15,000 × *g* for 5 min) and resuspended in 1 mL of the respective pre-warmed washing buffer for 30 min at 46°C.
2. After stringent washing, cells are recovered by centrifugation, washed in 1 mL PBS, resuspended in 20–50 μL PBS or water, placed on ice, and analyzed within 3 h.
3. For each sample, 10–50 μL of hybridized cell suspension is spotted onto pre-cleaned microscopic slides and dried in the dark at 37°C for 20 min (see Note 19).
4. Slides are mounted with a drop of Vectashield solution and examined with an epifluorescence microscope and a filter set for the fluorocrome Cy3 (see Note 20).
5. Microphotographs are taken using a digital camera and edited using standard software (see Note 11).
6. Positive hybridization signals are determined as bright orange fluorescent yeast cells (when probes are labeled with Cy3) (Fig. 1) (see Note 21).

4. Notes

1. Other international culture collections where this strain is available are *Centraalbureau voor Schimmelcultures*, Utrecht, The Netherlands (with the number CBS 132^{T}), *American Type Culture Collection*, Manassas, Virginia, USA (with the number ATCC 32045), and *ARS Culture Collection*, Peoria, Illinois, USA (with the number NRRL Y-2534). Other strains correctly identified as *Cryptococcus neoformans* can also be used for FISH assays.
2. Samples are received for routine conventional testing at the Institute of Hygiene and Tropical Medicine, Lisbon. Samples were previously examined for the presence of encapsulated budding yeasts by India ink preparation using microscopy and for the presence of cryptococcal polysaccharide antigen by latex agglutination test (Pastorex Crypto-PIus Biorad) and by culture on Sabouraud glucose agar (CSF samples), according to standard procedures (5).
3. A DNA probe for FISH should be fully complementary only to the respective target sequence and have more than one mismatch with the homologous region of non-target species (mismatches should preferably be situated at a central position). However, the occurrence of cross-reactions with non-target species is best prevented if three or more mismatches are present. The G+C content of 18–24 base oligonucleotide probes

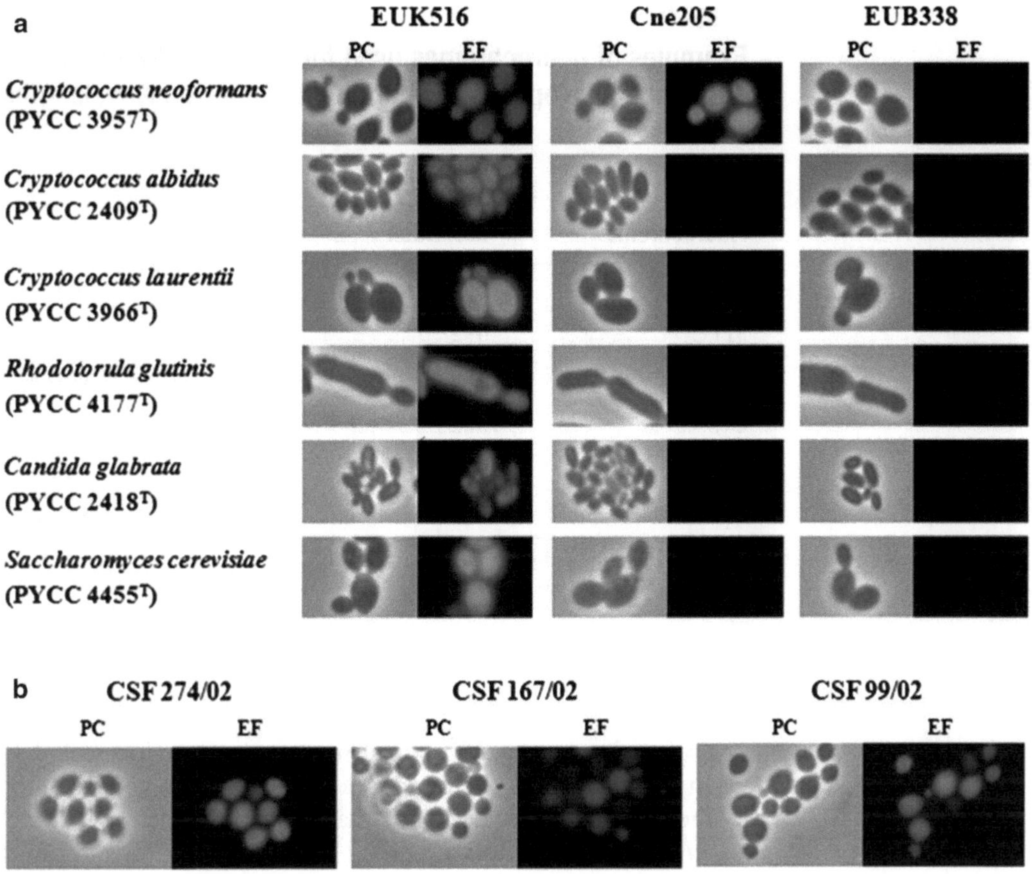

Fig. 1. Grayscale photomicrographs in phase contrast (PC) and epifluorescence (EF) of typical results of FISH assays. (**a**) Specificity tests with EUK516, Cne205, and EUB338 probes using strain PYCC 3957^{T} (*Cryptococcus neoformans*) and other non-target yeast species. (**b**) Examples of FISH assays using Cne205 probe with CSF samples. All *Cryptococcus neoformans* cultures and positive CSF samples should yield strong fluorescent signals with both the universal EUK516 and Cne205 probes. None of the yeast cells should yield any fluorescent signal with the bacterial EUB338 probe. Non-*Cryptococcus neoformans* yeast cells subjected to the same FISH protocol should render negative to both EUB338 and Cne205 but positive with EUK516.

should usually range between 50 and 60% (theoretical melting temperatures preferentially ranging from 46 to 60°C, according to the simple formula $T_m = 4 \times GC + 2 \times AT$). Smaller probes can penetrate cells more easily but may be also less specific. The fluorochromes most commonly used to label FISH probes are derived from fluorescein, rhodamine, and cyanine dyes such as Cy3 (Table 2). The latter have shown a superior performance, combining high fluorescence intensity with good stability. The fluorochrome molecules are usually linked to the 5′ end of oligonucleotide probes during synthesis or, alternatively, to the 3′ end by the action of a terminal transferase (1, 2). Not all sites within the rRNA molecules are readily accessible for FISH

Table 2
Examples of fluorochromes used for labeling FISH probes and respective optimal excitation and emission wavelengths

Fluorochrome	Wavelength (nm) Excitation	Wavelength (nm) Emission	Color
AMCA[a]	351	450	Blue
FITC[b]	492	528	Green
Fluo X®	488	520	Green
Cy3®	552	570	Orange
Texas Red®	578	600	Red
Cy5®	651	674	Infrared

[a] *AMCA* 7-amino-4-methylcoumarin-3-acetic acid
[b] *FITC* fluorescein isothiocyanate

probes due to, e.g., ribosomal structural constraints (12, 13). In this method probe target selection took into consideration the 26S rRNA accessibility map previously determined for *Saccharomyces cerevisiae* (13), regarding the hybridization efficiency of FISH probes to their respective rRNA targets. Similar rRNA accessibility maps are available for other organisms (12, 16, 17). The use of these color-coded maps of model organisms facilitates the selection of probe target sites promising high signal intensities due to the high evolutionary conservation of the ribosome among related organisms.

4. A 0.5 μmol synthesis scale of fluorescently labeled oligonucleotide probes is commercially available and provides enough probe for several thousand reactions. Prepare small aliquots of probe working solutions (up to 100 μL). The continuous freeze and thaw of the probe working solutions may cause the appearance of bright fluorescent artifacts under the microscope, leading also to faint specific hybridization signals and high background.
5. Use a mask for weighing PFA and prepare this solution under the hood, since this reagent causes irritation if inhaled. Ideally, this solution should be freshly prepared from paraformaldehyde, being usable for up to 1 week (kept at 4°C). Store the solution at –20°C for longer periods (up to 1 year).
6. Use a mask for weighing SDS since this reagent causes irritation if inhaled. Dissolve with stirring and a gentle warming.

7. DNA probes require the optimization of the respective stringent hybridization conditions case by case (such as the optimal formamide concentration in the hybridization buffer). Conditions reported here were previously optimized for the universal probes EUK516 and EUB338 (15) and Cne205 (5). The adequate hybridization stringency usually corresponds to the highest formamide concentration that does not result in loss of fluorescence intensity of the target cells and for which probe hybridization to the non-target organisms no longer occur.
8. Hybridization buffer should be freshly prepared for FISH assays from the respective constituent stock solutions. Wear suitable protective clothing and gloves for manipulating formamide, since this reagent is very toxic and irritating to eyes, respiratory system, and skin. Always use fresh deionized formamide, since this reagent is easily degraded, leading to unspecific FISH signals (take an aliquot from the stock and use it up within a week; fresh formamide should be colorless). SDS solution should be added last to avoid precipitation. If SDS stock solution is precipitated, warm before use. SDS denatures the native ribosome structure leading to an increased accessibility of the rRNA target sites to the complementary DNA probes. The absence of SDS in hybridization buffer may lead to a complete absence of fluorescence signals.
9. EDTA will dissolve at pH 8.0.
10. The stringency of the washing buffer is regulated by adjusting the salt concentration and EDTA, which avoids the use of excess amounts of formamide. Lowering the salt concentration in buffer encourages dissociation of mismatched heteroduplexes, yielding higher washing stringencies. SDS solution should be added last to avoid precipitation.
11. The described protocol was performed with an Olympus BX50 microscope, fitted for epifluorescence microscopy with either an U-ULH100W mercury high pressure bulb and an U-MA1007 filter set for the fluorocrome Cy3 (Olympus). Microphotographs were taken using a digital camera (Olympus C3030-Z00M) and edited using standard software (Adobe Photoshop 6.0, Adobe).
12. Metabolically active cells should be collected in the exponential growth phase. Fluorescence hybridization signals decrease significantly when the culture enters the stationary growth phase.
13. The fixation of the sample is one of the most critical steps in the protocol. The aim is to immobilize the cells facilitating the penetration of the probe and protecting ribosomes from degradation by endogenous RNAses keeping the cells mor-

phologically intact. The setting of fixation conditions can vary depending on the organisms and the type of samples. Standard fixatives are aldehydes and alcohols. Some fixation conditions can cause cell lysis, which can be assessed by comparing the number of cells before and after the fixation step with the aid of a hemacytometer (losses ranging from 10 to 20% in the number of cells are acceptable). Fixation can also cause changes in cell size, which can be evaluated under a microscope. Good results are usually achieved for yeast cell fixation using concentrations between 1 and 4% formaldehyde for 2–16 h at 4°C. After the fixation a dehydration series of 50, 80, and 96% ethanol may help to permeabilize cells for FISH, albeit this step was not considered necessary for *Cryptococcus neoformans.* If cells are over-fixed in PFA by incubation for longer periods, the formation of excessive cross-links between cell wall components may render cells completely impermeable to DNA probes. In our conditions, fixation of *Cryptococcus neoformans* cells using ethanol yielded very faint fluorescence signals after hybridization with EUK516 and Cne205 probes and therefore PFA fixation was considered better for these cells.

14. Adjust the volume of PBS and cold absolute ethanol when resuspending cell pellets in order to obtain concentrations around 10^6 cells/μL. Fixed cells can be maintained at −20°C for several months without apparent loss of FISH performances.
15. Hybridization experiments using unfixed *Cryptococcus neoformans* cells (taken directly from the culture) produced identical results to those obtained with PFA fixed cells, with fluorescence signals of similar intensities. These results indicate that the fixation step may be unnecessary for the direct analysis of biological samples, which can shorten the response time and avoid cell losses by diminishing centrifugation and washing steps.
16. Each strain is tested with the *Cryptococcus neoformans* specific probe (Cne205) and with the EUK516 and EUB338 probes, universal for eukaryotic and bacterial organisms, respectively.
17. In this chapter and in a previous publication (5) we describe FISH assays using cell suspensions. FISH reactions are also usually performed on cell smears fixed on microscopic glass slides, which are embedded in hybridization buffer and incubated in a humid chamber (18). Hybridization and washing requires precise temperature control to prevent nonspecific hybridization at low temperatures and loss of fully complementary probes at higher temperatures. Hybridization stringency is usually adjusted by varying the concentration of formamide in hybridization buffer, which allows the use of lower operating temperatures for performing FISH assays (an increase of 0.5% of formamide concentration in the hybridiza-

tion buffer is equivalent to an increase of 1°C in incubation temperature). However, hybridization stringency may also be adjusted by temperature rather than by the chemical composition of buffers. Temperature and time of hybridization and washing steps have to be optimized for each probe.

18. CSF samples were tested using only Cne205 probe since the availability of sufficient sample volumes for testing additional probes is limiting but in an ideal situation would be tested with all probes to confirm result.
19. Use well cleaned microscopic slides in order to avoid fluorescence artifacts. The simplest way of efficiently cleaning microscope slides is to soak the slides in ethanol and wipe clean with a lint-free cloth.
20. Wash skin with water if contact occurs with the mounting medium since it can cause skin irritation. Air-dried preparations, as well as mounted slides, can be stored in the dark at -20°C for several days without substantial loss of probe fluorescence.
21. Autofluorescence was not a problem when performing FISH with *Cryptococcus neoformans* cells, although some background fluorescence was observed in some preparations probably due to nonspecific probe binding. The addition of 0.2% bovine serum albumin or 100 ng/μL salmon sperm DNA to the hybridization buffer was not successful for inhibiting the occurrence of that background fluorescence. CSF, the biological sample most commonly received in clinical microbiology laboratories for the diagnosis of cryptococcosis, showed no autofluorescence in the conditions tested and contains no inhibitors of FISH reactions.

References

1. Amann R, Ludwig W, Schleifer K-H (1995) Phylogenetic identification and in situ detection of individual microbial cells without cultivation. Microbiol Rev 59:143–169
2. Moter A, Göbel UB (2000) Fluorescence in situ hybridization (FISH) for direct visualization of microorganisms. J Microbiol Methods 41:85–112
3. Zwirglmaier K (2005) Fluorescence in situ hybridisation (FISH)—the next generation. FEMS Microbiol Lett 246:151–158
4. Alexander BD, Ashley ED, Reller LB et al (2006) Cost savings with implementation of PNA FISH testing for identification of *Candida albicans* in blood cultures. Diagn Microbiol Infect Dis 54:277–282
5. Martins ML, Ferreira AS, Sampaio A et al (2010) Direct and specific identification of *Cryptococcus neoformans* in biological samples using fluorescently-labelled DNA probes. Eur J Clin Microbiol Infect Dis 29:571–576
6. Stender H, Kurtzman C, Hyldig-Nielsen JJ et al (2001) Identification of *Dekkera bruxellensis* (*Brettanomyces*) from wine by fluorescence in situ hybridization using peptide nucleic acid probes. Appl Environ Microbiol 67:938–941
7. Tavares S, Inácio J, Fonseca Á et al (2004) Direct detection of *Taphrina deformans* on peach trees using molecular methods. Eur J Plant Pathol 110:973–982
8. Xufre A, Albergaria H, Inácio J et al (2006) Application of fluorescence in situ hybridisation (FISH) to the analysis of yeast population dynamics in winery and laboratory grape must

fermentations. Int J Food Microbiol 108: 376–384

9. Andrews JH, Spear RN, Nordheim EV (2002) Population biology of *Aureobasidium pullulans* on apple leaf surfaces. Can J Microbiol 48:500–513
10. Li S, Spear RN, Andrews JH (1997) Quantitative fluorescence in situ hybridization of *Aureobasidium pullulans* on microscopic slides and leaf surfaces. Appl Environ Microbiol 63:3261–3267
11. Inácio J, Ludwig W, Spencer-Martins I et al (2010) Assessment of phylloplane yeasts on selected Mediterranean plants by fluorescence in situ hybridisation (FISH) with group- and species-specific oligonucleotide probes. FEMS Microbiol Ecol 71:61–72
12. Behrens S, Rühland C, Inácio J et al (2003) In situ accessibility of small subunit ribosomal RNA of members of the domains bacteria, archaea and eucarya to Cy3-labeled oligonucleotide probes. Appl Environ Microbiol 69:1748–1758
13. Inácio J, Behrens S, Fuchs BM et al (2003) In situ accessibility of *Saccharomyces cerevisiae* 26S rRNA to Cy3-labelled oligonucleotide probes comprising the D1 and D2 domains. Appl Environ Microbiol 69:2899–2905
14. Casadevall A, Perfect JR (1998) *Cryptococcus neoformans*. ASM Press Books, Washington, DC
15. Amann R, Krumholz L, Stahl DA (1990) Fluorescent-oligonucleotide probing of whole cells for determinative, phylogenetic, and environmental studies in microbiology. J Bacteriol 172:762–770
16. Fuchs BM, Wallner G, Beisker W et al (1998) Flow cytometric analysis of the in situ accessibility of *Escherichia coli* 16S rRNA for fluorescently labeled oligonucleotide probes. Appl Environ Microbiol 64:4973–4982
17. Fuchs BM, Syutsubo K, Ludwig W et al (2001) In Situ accessibility of *Escherichia coli* 23S rRNA to fluorescently labeled oligonucleotide probes. Appl Environ Microbiol 67:961–968
18. Tavares A, Inácio J, Melo-Cristino J et al (2008) Use of fluorescence in situ hybridization for rapid identification of staphylococci in blood culture samples collected in a Portuguese hospital. J Clin Microbiol 46:3097–3100

Chapter 6

Quantitative Detection of *Aspergillus* spp. by Real-Time Nucleic Acid Sequence-Based Amplification

Yanan Zhao and David S. Perlin

Abstract

Rapid and quantitative detection of *Aspergillus* from clinical samples may facilitate an early diagnosis of invasive pulmonary aspergillosis (IPA). As nucleic acid-based detection is a viable option, we demonstrate that *Aspergillus* burdens can be rapidly and accurately detected by a novel real-time nucleic acid assay other than qPCR by using the combination of nucleic acid sequence-based amplification (NASBA) and the molecular beacon (MB) technology. Here, we detail a real-time NASBA assay to determine quantitative *Aspergillus* burdens in lungs and bronchoalveolar lavage (BAL) fluids of rats with experimental IPA.

Key words: Real time, Nucleic acid sequence-based amplification, Molecular beacon, *Aspergillus* burden, Quantitative

1. Introduction

Invasive pulmonary aspergillosis (IPA) is a major cause of morbidity and mortality in immunocompromised patients. Therapeutic success depends on early diagnosis and appropriate initiation of antifungal therapy (1). Nevertheless, early diagnosis of IPA continues to be problematic for this population due to the lack of rapid, sensitive, and specific diagnostic methods. The combination of galactomannan (GM) detection and standard clinical, radiological, and histological examinations has been an important advance in the detection of IPA. However, GM testing has false positives and decreased sensitivity in patients under antifungal prophylaxis (2–4), which highlights a

Louise O'Connor and Barry Glynn (eds.), *Fungal Diagnostics: Methods and Protocols*, Methods in Molecular Biology, vol. 968,
DOI 10.1007/978-1-62703-257-5_6, © Springer Science+Business Media New York 2013

critical need for more sensitive, specific, and reliable diagnostic techniques. Molecular diagnostics have emerged as a promising method for the diagnosis of a variety of infectious diseases, including fungal infections (5). Real-time quantitative PCR (qPCR) has been extensively studied and explored as a tool for detecting and identifying *Aspergillus* and other pathogenic fungi in clinical samples (6–10), although the PCR technique still hasn't been approved for defining IA because of lack of standardization and the absence of a commercially available system (11, 12).

Nucleic acid sequence-based amplification (NASBA) is an isothermal amplification process that specifically amplifies RNA even in the presence of genomic DNA (13). The amplification of NASBA is highly robust yielding $>10^{12}$ amplicons in 30 min (14). The amplified single-stranded RNA can be detected easily in real time by molecular beacon (MB) probes, which are self-reporting, hairpin structured, single-stranded nucleic acid probes that brightly fluoresce when bound to their targets (15). As an alternative to PCR for molecular detection, real-time NASBA has been widely used for detecting RNA viruses such as enterovirus and HIV (16–18), certain microbial pathogens including *Legionella* species, *Vibrio cholerae* (19, 20), and pathogens from food and environmental samples (21, 22). To date, only studies of small scope have been reported using NASBA to detect *Aspergillus* species (23, 24), although a good diagnostic value was demonstrated by these studies. To extend the applicability of real-time NASBA assay to the quantification of the fungal load in order to monitor the progress of *Aspergillus* infection, we established a pan-*Aspergillus* real-time NASBA assay and validated with an experimented IPA model (25). We chose to develop a pan-*Aspergillus* assay to gain complete capture of *Aspergillus* species, even though *A. fumigatus* remains the most abundant invasive species.

Quantification of nucleic acids by real-time NASBA is achieved by correlating an increase in probe fluorescence to the amount of starting template. The point at which the fluorescence signal rises above the cutoff value (20% higher than the end-point signal from no template control) is called the time to positivity (TTP) (Fig. 1a). The dynamic linear quantification range of different probes varies and has to be determined by a standard curve method (Fig. 1b), which shows a linear regression relationship between the log of the amount of starting template and the corresponding TTP value. To achieve reliable *Aspergillus* burden estimation, we have outlined several essential steps which include: (1) total nucleic acid extraction from lung tissue and bronchoalveolar lavage (BAL) fluids, (2) external standard

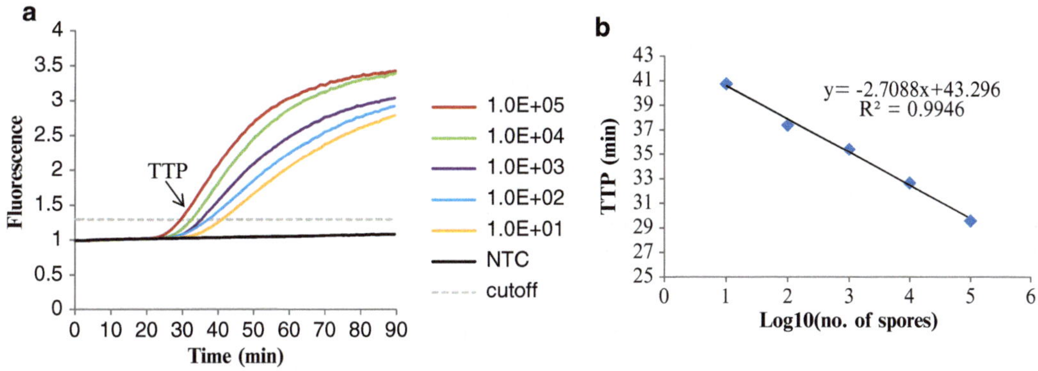

Fig. 1. (**a**) Amplification curves of the real-time pan-*Aspergillus* NASBA assay are produced using total nucleic acids extracted from serially diluted *A. fumigatus* conidia with 10^5 to 10^1 spores. The TTP value increases as the amount of template initially present in the reaction decreases. (**b**) The standard curve is constructed by plotting the log of starting amount against the corresponding TTP value. *Aspergillus* burden estimation is viable by simply applying TTP values from each sample to this linear regression equation.

series preparation, (3) molecular beacon purification and characterization, (4) real-time NASBA assay, and (5) standard curve and burden quantification.

2. Materials

Prepare all solutions using nuclease-free water and DNase, RNase-free molecular grade reagents. All waste materials should be disposed of in accordance with local waste disposal regulations.

2.1. *Aspergillus fumigatus* Standards

1. Potato dextrose agar (PDA) slant: suspend 39 g of PDA medium in 1 L of purified water. Heat and stir frequently to completely dissolve the medium. Autoclave at 121°C for 15 min. Pour the medium into a 25 ml Flat Base Polystyrene Screw Cap Tube and keep the tube in a slanting position until the agar solidifies.
2. Sterile 0.1% Tween 20 solution.
3. Disposable hemocytometer and an optical microscope with a 200× objective lens.

2.2. Total Nucleic Acid Extraction

1. RNA*later* solution (Ambion, Austin, TX).
2. NucliSENS lysis buffer (bioMérieux, Durham, NC) (see Note 1).
3. Proteinase K solution.
4. Lysing matrix D (MP, Biomedicals, Inc., Solon, OH) (see Note 2).

5. FastPrep®-24 instrument (MP, Biomedicals, Inc., Solon, OH).
6. EasyMAG® wash buffer 1, 2, 3 and EasyMAG® magnetic silica solution (bioMérieux, Durham, NC).

2.3. Molecular Beacon Purification and Melting Curve Analysis

1. Molecular beacon probes were labeled at the 5′-end with fluorophore 5-carboxyfluorescein (FAM) and quencher dabcyl (4-(4-dimethylaminophenyl) diazenylbenzoic acid) at the 3′-end.
2. Beckman Coulter System Gold® high-pressure liquid chromatography (HPLC) systems (Beckman Coulter, Fullerton, CA).
3. Buffer A: 1% triethylamine (TEA), adjust pH to 6.5 by Glacial acetic acid. Filtrate through a NALGENE or equivalent bottle filter (0.45 μm nylon filter).
4. Buffer B: 1% TEA, 75% acetonitrile, adjust pH to 6.5 by Glacial acetic acid. Filtrate through a NALGENE bottle filter (0.45 μm nylon filter) with 3–4 extra sterilized filter paper.
5. 3 M Sodium acetate (NaOAc).
6. 100% ice-cold ethanol.
7. Tris–EDTA (TE) buffer, pH 8.0.
8. Melting curve buffer: 1× PCR buffer with 4 mM $MgCl_2$.
9. Target oligonucleotide (100 μM): short oligonucleotide that is perfectly complementary to the molecular beacon probe sequence.
10. Stratagene Mx4000 or Mx3005P real-time instrument (Agilent Technologies, Santa Clara, CA).

2.4. Real-Time NASBA

1. NucliSENS EasyQ® Basic kit v.2 (bioMèrieux).
2. Primers detecting *Aspergillus* spp. targets 28S ribosomal RNA. The P1 primer is 5′-AATTCTAATAC GACTCACTATA GGGGA GAATCCACATCCAGGTGC-3′, and P2 primer is 5′-CAGCAGTTGGACATGGGTTA-3′ (see Note 3). Prepare a 20 μM working solution of each primer and store at −20°C until use.
3. A molecular beacon is designed for this pan-*Aspergillus* assay. The sequence is 5′-FAM-CGCGAGTGCGCCGTGTGCCGAA ATCGCG-Dabcyl-3′. The working solution is 10 μM and stored at −20°C until use.
4. 0.2 ml thin-wall PCR 8-tube strip (free of RNase, DNase, DNA, and PCR inhibitors).
5. NucliSENS easyQ® system (bioMérieux).

3. Methods

3.1. TNA Extraction from Lung Tissue and BAL Fluids of Rats with IPA

1. Thaw frozen lung homogenates (1 g lung tissue per sample) in RNA*later* or BAL samples (1 ml BAL+2 ml RNA*later*) at room temperature and centrifuge at 16,000 × *g* for 10 min. Discard supernatant and resuspend the pellet in 1 ml of NucliSENS lysis buffer with proteinase K (100 μg/ml).
2. Transfer the entire solution into the lysing matrix D tube and vigorously vortex in a FastPrep instrument at speed level 6 for 45 s twice (see Note 4). Incubate at 55°C for 1 h (see Note 5).
3. Centrifuge cell lysates at 16,000 × *g* for 10 min. In the meantime, add 2 ml of NucliSENS lysis buffer into each well of the NucliSENS easyMAG® sample cartridge and dilute the magnetic silica twofold. Transfer the supernatant into the easyMAG® sample cartridge and mix with lysis buffer. Mix 140 μl of the diluted magnetic silica solution with each sample.
4. On easyMAG® system, input sample record and choose the extraction protocol "specific B" using the easyMAG® software. The final elution volume option increases from 40 to 110 μl at a 5 μl step. Choose the volume which best fits your experiment.
5. Store the TNA samples at -80°C until use.

3.2. External Standards Preparation and TNA Extraction

1. Grow *Aspergillus fumigatus* wild-type strain R21 on a PDA slant at 37°C for 3 days.
2. Add 2 ml of the 0.1% Tween20 into the slant. Collect and transfer the conidial suspension in a clean 2 ml tube. Vortex until the conidia is evenly distributed in the suspension.
3. Prepare tenfold serial dilutions of the above conidia suspension in 0.1% Tween20. Load 100- and 1,000-fold dilution on the hemocytometer to count the conidia (see Note 6).
4. Based on the cell counts acquired in step 3 prepare 1 ml of 5×10^6 CFU/ml conidia suspension in sterile 0.1% Tween20. Prepare further tenfold serial dilutions down to 5 CFU/ml.
5. Mix a 200 μl aliquot of each dilution with 800 μl of easyMAG lysis buffer (with 100 μg/ml of Proteinase K) and vortex on a FastPrep® instrument as described for lung tissue and BAL samples. The remaining steps of the extraction are the same.

3.3. Molecular Beacon Purification and Melting Curve Analysis

1. Resuspend the crude beacon in 500 μl of Buffer A. Incubate at room temperature for 5 min.
2. Spin product in 0.22 μm filter on a microcentrifuge at maximum speed for 3 min.

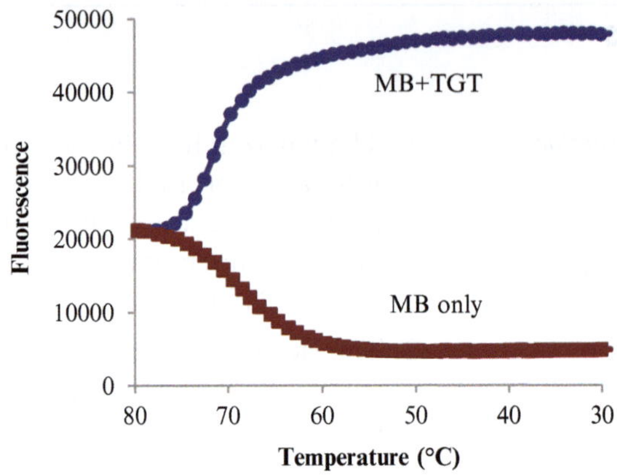

Fig. 2. Thermal denaturation profile shows a wide annealing range of the molecular beacon 28SPA.

3. Collect the filtrate and purify the crude molecular beacon by HPLC through a C-18 reverse-phase column, using a linear elution gradient of 20–80% buffer B that forms over 25 min at a flow rate of 1 ml/min. Monitor the absorption at 260 nm for nucleic acid and at 495 nm for FAM, and collect the fractions with peaks at both wavelengths, as described: http://www.molecular-beacons.com/MB_SC_protocol.html.
4. Precipitate the collected fractions with 2.5 volumes of cold 100% ethanol and 1/10 volume of 3 M NaOAc at –20°C for at least 2 h or overnight (see Note 7).
5. Spin in a microcentrifuge at maximum speed for 20 min at 4°C.
6. Aspirate the supernatant and dry the pellet.
7. Resuspend the pellet in 50 μl of TE buffer.
8. Measure the concentration of the purified molecular beacon. Prepare a 10 μM working solution and store at –20°C.
9. Prepare two 50 μl reactions that contain 200 nM of the molecular beacon in 1× melting curve buffer. Add a fivefold molar excess of the target oligonucleotide to one tube and add only buffer to another tube.
10. On a Stratagene Mx4000 or Mx3005P real-time instrument, run the molecular beacon melting curve program to determine the fluorescence of each solution as a function of temperature. Decrease the temperature of these tubes from 80 to 25°C in 1°C steps, with each hold lasting 30 s, while monitoring the fluorescence during each hold (Fig. 2).

3.4. Real-Time NASBA Assay

1. Prepare enzyme by adding 45 μl of enzyme diluent to enzyme accusphere (red cap), incubate at room temperature for 20 min before use (see Note 8). Each sphere is intended for 8

reactions, for additional reactions pool spheres and add the appropriate amount of diluent. Unused enzyme may be stored at −80°C for up to 2 weeks with 1 freeze–thaw cycle.

2. To create a primer/beacon mixture, add equal parts of primer P1, primer P2, molecular beacon, and Nuclease-free water, e.g.: mix 2 μl of each for a total of 8 μl for 8 reactions.
3. Prepare reagent mixture as follows (for 8 reactions): In a clean tube, add 64 μl of reagent diluent plus 9.6 μl of water and 14.4 μl of KCl (final concentration is 90 mM). Add 8 μl of primer/beacon mix. Add reagent accusphere (blue cap) to tube and vortex immediately (see Note 9).
4. Aliquot 10 μl of reagent mixture to each well in 8 tube strip.
5. Add 5 μl of TNA sample to each well in 8 tube strip.
6. Incubate strips in easyQ incubator for 2 min at 65°C and 2 min at 41°C (program 1).
7. While strips are incubating, add 5 μl enzyme into 8 strip caps.
8. After incubation, secure caps on tubes. Spin briefly in microcentrifuge, flick tube to mix thoroughly, and spin again. Move strips to easyQ analyzer (see Note 10).
9. Open "NucliSens® EasyQ Sample Login," enter sample ID and plate setup, and save the file in the default folder. Open "NucliSens® EasyQ Analysis Software," go to File, and open the file saved above. Select Edit, change the run time to 90 min (see Note 11). Click the Runs button located on the left margin, and place strips onto the NucliSens® EasyQ plate. Confirm that the temperature is 41°C, then click Start (see Note 12).
10. When run is complete (00:00 time is left), click "Out" to move tray to out position. Remove tube strips from the NucliSens® EasyQ Analyzer and discard properly. Go to "NucliSens® EasyQ Data Analyzer" to analyze the data.

3.5. Data Analysis and Burden Quantification

1. Open "NucliSens® EasyQ Data Analyzer." Click the "Start Import of Measurement Data" button, select corresponding file and open.
2. View general tab, calculated data tab, or raw graphic data tabs to view results.
3. Set the threshold as 20% higher than the endpoint signal from the extraction negative control and determine the TTP for each sample on the calculated data tab, then export the data into a new excel spreadsheet.
4. Using the log number of conidia in the external standard and the paired TTP data, draw a standard curve and evaluate the linearity by calculating the R square and the linear equation and determine the best linear quantification range (see Fig. 1b).

5. Calculate the conidia equivalents (CE) for each TNA sample extracted from lungs and BAL fluids by placing the TTP values on the standard curve (see Note 13).
6. Estimate the fungal burden in lung tissue and BAL fluids after accounting for the dilution factor.

4. Notes

1. NucliSENS lysis buffer can be stored at an ambient temperature from 2 to 30°C before opening but should be stored at 2–8°C after opening. Make sure to equilibrate the buffer at room temperature for at least 30 min before use if it is stored at lower temperature.
2. We choose lysing matrix D in this protocol because other matrix tubes provided by MP contain smaller beads, which tend to clog the narrow ends of the disposable tips used on the easyMAG® instrument and may result in an extraction failure.
3. The Pan-*Aspergillus* real-time NASBA assay was designed for detecting *Aspergillus* spp. and showed a high degree of specificity >95%. However, due to the fact that *Penicillium chrysogenum* is 99% homologous to *Aspergillus* spp. in the amplifying region, the cross-reaction from this species would be expected. Other species like the AIDS pathogen *P. marneffei* (26) have the potential to cross-react. This assay did not cross-react with other fungal and bacterial species tested in our lab.
4. Place samples on ice between each FastPrep cycle to rapidly cool down the temperature inside of the tube. When finishing the FastPrep step open cap to degas before the incubation. Because the foam will be produced during the FastPrep, cooling and degassing will help to reduce the volume.
5. The incubation time can be increased if any visible tissue chunk still exists after 1 h incubation. Additional time of digestion and the shaking platform help thorough digestion.
6. Depending on the density of the original conidia suspension, greater dilutions may be needed for accurate counting.
7. At least 2 h incubation time at −20°C is required. Less time for precipitation will reduce the final yield of the purified beacon.
8. Do not vortex the enzyme. Enzyme must be used within 1 h of reconstitution.
9. If more than 8 reactions are desired, pool reagent diluent, water, KCl, and primer/beacon mix. Add appropriate number of reagent accuspheres to tube and vortex immediately. If less

than 8 reactions are desired, add 64 μl reagent diluent directly to reagent accusphere and vortex immediately. Once sphere is in solution, reagent may be aliquoted and stored at -70°C for up to 2 weeks with 1 freeze–thaw. Proceed with addition of water, KCl, and primer/beacon mix. Do not freeze after addition of water, KCl, and primer/beacon mix.

10. These steps should be performed quickly to keep temperature at 41°C.
11. Make sure that the sample ID, plate setup, and running profile are ready before proceeding with NASBA assay setup. Do not leave them until step 8 is finished, otherwise it won't be possible to get the instrument ready without miscalculating the TTP.
12. Do not close the window of "NucliSens® EasyQ Analysis Software" before NASBA run is completed (data will be lost).
13. The CE calculation can only be made within the linear quantification range of the standard curve. Be careful when the TTP value falls outside of the range. The externalization is not recommended. Because the amplification efficiency, which is reflected by the slope of the equation, is different from what is determined for the linear quantification range when the amount of template is too low or too high, the application of the equation on outliers will bias the real fungal burden. Thus, leave the CE estimation as such (<lower linear quantification limit or > higher linear quantification limit) when you have outliers.

Acknowledgments

This work was supported by NIH grant AI066561 to D.S.P.

References

1. Denning DW (1998) Invasive aspergillosis. Clin Infect Dis 26:781–803, Quiz 804–785
2. Marr KA, Laverdiere M, Gugel A et al (2005) Antifungal therapy decreases sensitivity of the Aspergillus galactomannan enzyme immunoassay. Clin Infect Dis 40:1762–1769
3. Pinel C, Fricker-Hidalgo H, Lebeau B et al (2003) Detection of circulating Aspergillus fumigatus galactomannan: value and limits of the Platelia test for diagnosing invasive aspergillosis. J Clin Microbiol 41:2184–2186
4. Sulahian A, Boutboul F, Ribaud P et al (2001) Value of antigen detection using an enzyme immunoassay in the diagnosis and prediction of invasive aspergillosis in two adult and pediatric hematology units during a 4-year prospective study. Cancer 91:311–318
5. Hope WW, Walsh TJ, Denning DW (2005) Laboratory diagnosis of invasive aspergillosis. Lancet Infect Dis 5:609–622
6. Imhof A, Schaer C, Schoedon G et al (2003) Rapid detection of pathogenic fungi from clinical specimens using LightCycler real-time fluorescence PCR. Eur J Clin Microbiol Infect Dis 22:558–560
7. Jordanides NE, Allan EK, McLintock LA et al (2005) A prospective study of real-time panfungal PCR for the early diagnosis of invasive fungal infection in haemato-oncology patients. Bone Marrow Transplant 35: 389–395
8. Spiess B, Buchheidt D, Baust C et al (2003) Development of a LightCycler PCR assay for detection and quantification of Aspergillus

fumigatus DNA in clinical samples from neutropenic patients. J Clin Microbiol 41: 1811–1818

9. Klingspor L, Jalal S (2006) Molecular detection and identification of Candida and Aspergillus spp. from clinical samples using real-time PCR. Clin Microbiol Infect 12:745–753
10. Kami M, Fukui T, Ogawa S et al (2001) Use of real-time PCR on blood samples for diagnosis of invasive aspergillosis. Clin Infect Dis 33: 1504–1512
11. Ascioglu S, Rex JH, de Pauw B et al (2002) Defining opportunistic invasive fungal infections in immunocompromised patients with cancer and hematopoietic stem cell transplants: an international consensus. Clin Infect Dis 34:7–14
12. Chen SC, Halliday CL, Meyer W (2002) A review of nucleic acid-based diagnostic tests for systemic mycoses with an emphasis on polymerase chain reaction-based assays. Med Mycol 40:333–357
13. Simpkins SA, Chan AB, Hays J et al (2000) An RNA transcription-based amplification technique (NASBA) for the detection of viable Salmonella enterica. Lett Appl Microbiol 30: 75–79
14. Compton J (1991) Nucleic acid sequence-based amplification. Nature 350:91–92
15. Tyagi S, Kramer FR (1996) Molecular beacons: probes that fluoresce upon hybridization. Nat Biotechnol 14:303–308
16. Rutjes SA, Italiaander R, van den Berg HH et al (2005) Isolation and detection of enterovirus RNA from large-volume water samples by using the NucliSens miniMAG system and real-time nucleic acid sequence-based amplification. Appl Environ Microbiol 71:3734–3740
17. Landry ML, Garner R, Ferguson D (2005) Real-time nucleic acid sequence-based amplification using molecular beacons for detection of enterovirus RNA in clinical specimens. J Clin Microbiol 43:3136–3139
18. McClernon DR, Vavro C, St Clair M (2006) Evaluation of a real-time nucleic acid sequence-based amplification assay using molecular beacons for detection of human immunodeficiency virus type 1. J Clin Microbiol 44:2280–2282
19. Fykse EM, Skogan G, Davies W et al (2007) Detection of Vibrio cholerae by real-time nucleic acid sequence-based amplification. Appl Environ Microbiol 73:1457–1466
20. Loens K, Beck T, Goossens H et al (2006) Development of conventional and real-time NASBA for the detection of Legionella species in respiratory specimens. J Microbiol Methods 67:408–415
21. Nadal A, Coll A, Cook N et al (2007) A molecular beacon-based real time NASBA assay for detection of Listeria monocytogenes in food products: role of target mRNA secondary structure on NASBA design. J Microbiol Methods 68:623–632
22. Churruca E, Girbau C, Martinez I et al (2007) Detection of Campylobacter jejuni and Campylobacter coli in chicken meat samples by real-time nucleic acid sequence-based amplification with molecular beacons. Int J Food Microbiol 117:85–90
23. Loeffler J, Hebart H, Cox P et al (2001) Nucleic acid sequence-based amplification of Aspergillus RNA in blood samples. J Clin Microbiol 39:1626–1629
24. Yoo JH, Choi SM, Lee DG et al (2007) Comparison of the real-time nucleic acid sequence-based amplification (RTi-NASBA) with conventional NASBA, and galactomannan assay for the diagnosis of invasive aspergillosis. J Korean Med Sci 22:672–676
25. Zhao Y, Park S, Warn P et al (2010) Detection of Aspergillus fumigatus in a rat model of invasive pulmonary aspergillosis by real-time nucleic acid sequence-based amplification. J Clin Microbiol 48:1378–1383
26. Vanittanakom N, Cooper CR Jr, Fisher MC et al (2006) Penicillium marneffei infection and recent advances in the epidemiology and molecular biology aspects. Clin Microbiol Rev 19:95–110

Chapter 7

Differentiation of Fungi Using Hybridization Probes on the LightCycler®

Stephan Fricke, Nadja Hilger, Christopher Oelkrug, Arne C. Rodloff, and Christian Fricke

Abstract

The polymerase chain reaction is a powerful molecular tool for the detection and analysis of very small amounts of DNA. Today, hybridization probes are often used in real-time PCR for more sensitive and specific detection of pathogens and for determination of gene regulation or mutation analysis instead of intercalating dyes like SYBR Green. Here, we describe how to generate suitable primers and hybridization probes for the specific detection of fungal DNA. Furthermore, we show the advantages of hybridization probes using the LightCycler-PCR for the detection of different *Candida* spp. and *Aspergillus* spp. in patient blood samples. The methods used to develop such PCR assays will also be presented in the following protocol.

Key words: Real-time PCR, Hybridization probes, Primer design, Databases, Aspergillus, Candida

1. Introduction

The most critical factor in medicine is the analysis of a wide range of parameters in small volume patient samples. Often immunosuppressed patients, e.g., after solid organ transplantation or hematopoietic stem cell transplantation, are affected by a number of complications like infections with bacteria, viruses, and especially fungi (*Candida* spp. or *Aspergillus* spp.) that also increase mortality (1–3). An early diagnosis of invasive fungal infections is essential for the initiation of specific antifungal therapies and to avoid the unnecessary administration of baseline therapy especially in cancer patients undergoing chemotherapy. Therefore, methods for an early detection of fungal infections are still warranted since current diagnostic tools lack sufficient sensitivity for the early diagnosis.

Louise O'Connor and Barry Glynn (eds.), *Fungal Diagnostics: Methods and Protocols*, Methods in Molecular Biology, vol. 968, DOI 10.1007/978-1-62703-257-5_7,

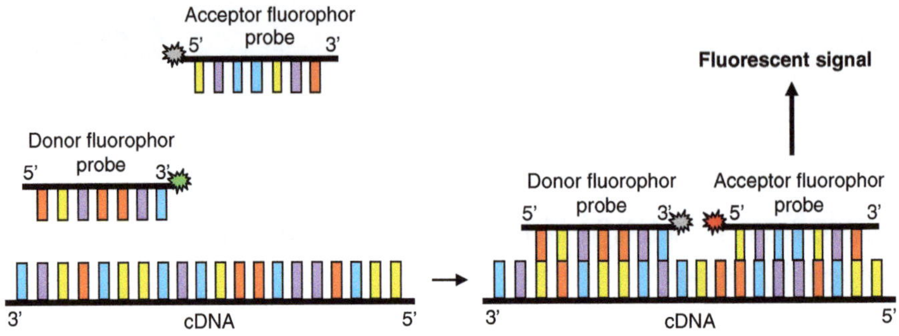

Fig. 1. Principle of FRET using hybridization probes.

Real-time PCR is a powerful and reliable tool for the detection of pathogens in serum, tissue, or blood and allows additional quantification of the amplified DNA in real-time by using fluorescent dyes (4, 5). The basic principle of such PCR assays is so-called fluorescence resonance energy transfer (FRET, Fig. 1). A fluorescent donor molecule transfers its energy to a nearby acceptor molecule by changing the fluorescence intensity. There are different types of fluorescent probes, e.g., TaqMan probes or hybridization probes. If the fluorescent donor probe is close to the acceptor probe and both hybridization probes bind to the target gene sequence (e.g., DNA of fungi), the acceptor probe emits a fluorescent signal proportional to the amplified target DNA. The generation and analysis of melting curves after DNA amplification allows the differentiation of *Aspergillus* spp. and *Candida* spp. (6).

We describe the development of primers and hybridization probes for detection of different *Candida* spp. and *Aspergillus* spp. in one sample.

2. Materials

2.1. Databases for Design of Primers and Hybridization Probes

1. Gene and sequence database: http://www.ncbi.nlm.nih.gov/guide/.
2. Database for intron and exon research: http://www.genome.ucsc.edu/.
3. Free online tool for design of primers, e.g., http://www.frodo.wi.mit.edu/primer3/.
4. Free online tool for proofing of primers: http://www.premierbiosoft.com/netprimer/index.html.
5. Roche UniversalProbeLibary: https://www.roche-applied-science.com/sis/rtpcr/upl/index.jsp.
6. Alignment software; VectorNTI (not free): http://www.invitrogen.com.

2.2. PCR

1. The LightCycler® 480 Real-Time-PCR System, LightCycler® 480 Instrument, and Data station and software (Roche Applied Science, Mannheim, Germany).
2. LightCycler® 480 Probes Master Kit (Roche Applied Science, Mannheim, Germany) for real-time PCR. Store at –20°C.
3. Fungal DNA (50 ng/µl): For isolation of DNA from *Candida* spp. and *Aspergillus* spp. the QiaAmp DNA Mini Kit (QIAGEN, Düsseldorf, Germany) or equivalent can be used according to the manufactures protocol (see Note 1). Store at –20°C.
4. 18SrRNA Primers (7) (100 pmol/µl): sense primer 5 µM (5′-ATT GGA GGG CAA GTC TGG TG-3′) and anti-sense primer 20 µM (5′-CCG ATC CCT AGT CGG CAT AG-3′) for amplification of 18SrRNA of fungi (TIB MOLBIOL, Berlin, Germany). Store at –20°C.
5. Hybridization probes (7) (1 nmol) for *Candida* spp.: (1) probe (5′-AGC CTT TCC TTC TGG GTA GCC ATT-3′) labeled with fluorescein (FL1) at 3′ end and (2) probe (5′-TGG CGA ACC AGG ACT TTT ACT TTG A-3′) labeled with LightCyler 640 Red (LC1) (TIB MOLBIOL, Berlin, Germany). Store at –20°C.
6. Hybridization probes (1 nmol) for differentiation between *Candida* spp. and *Aspergillus* spp., probe 1 (5′-CCA AGG ACG TTT TCA TTA ATC AAG AAC GA-3′) labeled with fluorescein (FL2) at 3′ end and probe 2 (5′-TTA GGG GAT CGA AGA TGA TCA GAT ACC-3′) (Fricke et al. unpublished data) labeled with LightCyler 640 Red (LC2). Store at –20°C.
7. Molecular grade water (free of DNAse and RNAse). Store at 4°C or room temperature.
8. LightCycler® 480 Multiwell Plate 96 and plate foil covers (LightCycler® 480 Sealing Foil, Roche Diagnostics, Roche Applied Science, Mannheim, Germany). Store at room temperature.

2.3. Gel Electrophoresis

1. Tris–Acetate–EDTA (TAE) buffer (50×) as running buffer for gel electrophoresis. The buffer should be diluted 1:50 for a 1× concentrated working solution. Store the TAE (50×) stock solution at 4°C.
2. High grade agarose for the separation of PCR fragments. Store at room temperature.
3. Blue/Orange 6× Loading Dye to determine the DNA concentration and to observe the progress of gel electrophoresis. Use 3 µl of the Loading Dye for 10 µl of the PCR sample. Store at 4°C.
4. 100 base pair (bp) DNA size marker (see Note 2). Store at 4°C.

5. Ethidium bromide solution 10 mg/ml for DNA staining of PCR fragments (see Note 3). Store at 4°C.
6. Comb for providing slots in the agarose gel.
7. Gel tray for casting the agarose gel.
8. Standard electrophoresis chamber for separating the amplified gene fragments according to their length.
9. Power pack as a power source for gel electrophoresis.
10. Image acquisition system for detection of PCR fragments under ultra violet (UV) light.
11. Microwave for preparing the agarose gel solution.

3. Methods

3.1. Finding the Target Gene Sequence for Fungal 18SrRNA

1. Open the online gene database NCBI (http://www.ncbi.nlm.nih.gov/guide/), which is the most used database for finding the target sequence. Put the name of the species in the search field, e.g., *Candida albicans* and the target gene 18SrRNA. A new window will be opened (see Note 4).
2. Click on "Nucleotide" and a new window will be opened with all matches of the *database*.
3. *Search* for the target gene "*Candida albicans* 18SrRNA gene" to see the information about the gene sequence.
4. Retrieve the DNA sequence and copy the sequence to a new work sheet or to available alignment software, e.g., VectorNTI (http://www.invitrogen.com), because the sequence will be important for the design of primers and hybridization probes.

3.2. Design of Primers

3.2.1. Selection of Primers

It is important that for real-time PCR the length of the amplified fragment should not exceed 200 bp (see Note 5).

1. There are some free online primer design tools that can be used, e.g., Primer3 (http://www.frodo.wi.mit.edu/primer3/)(8).
2. For Primer3 put your sequence in the search field (numbers in front of gene sequence can be ignored) and define the parameters. Most online tools for primer design have selected automatically the best primer standard conditions, e.g., G/C concentration of 40–60% or a length of 18–20 bp.
3. Select "Pick primers."
4. A window will appear with a collection of different primer pairs with information about their sequence, length of the amplified fragment, melting temperature (T_m), and position in the sequence (see Notes 6 and 7).

3.2.2. Verification of Primers

1. For quality control and verification of PCR primers, free online tools, e.g., Netprimer under http://www.premierbiosoft.com/netprimer/index.html (registration needed) or Beacon Designer™ Free Edition can be used. The generated primers should not form dimers or crossdimers (see Note 8).
2. Add the sequence of the sense and anti-sense primer in the respective field (Sequence) in 5′-3′ direction.
3. Select the button "Analyze" and information about melting temperature, molecular weight, hairpins, dimers, crossdimers, and palindromes will be generated.

3.3. Design of Hybridization Probes

Hybridization probes are short fluorescent labeled oligonucleotide sequences (about 20 bp), which specifically bind within the target sequence. Hybridization probes are more sensitive than common intercalating DNA dyes (e.g., SYBR green) because of a higher specificity (see Note 9). For the detection of different *Candida* spp., hybridization probes were designed because of their advantage in comparison to TaqMan probes (see Note 10). Thus, some manufacturers like Roche Applied Science developed an online database Roche UniversalProbeLibary (https://www.roche-applied-science.com/sis/rtpcr/upl/index.jsp) where primers and compatible hybridization probes can be found and bought for common species, e.g., human, mouse, drosophila, or arabidopsis thaliana. For other species, especially fungi, hybridization probes have to be designed manually by the same procedure described in Subheading 3.2 but some specific criteria are necessary as follows:

1. The sequence which is amplified by the designed primers is necessary (see Subheading 3.2, step 4) because hybridization probes have to bind within this sequence.
2. Both hybridization probes bind on the same strand. The distance between the two hybridization probes should be 1–5 bases (see Note 11).
3. The melting temperature (T_m) of the hybridization probes should be 5–10°C over the T_m of the collected primers but <80°C (see Notes 12 and 13).
4. The hybridization probes should not form dimers with the primers and should be phosphorylated at the 3′ so they cannot serve as primers.
5. The hybridization probe in the 5′-3′ direction should be labeled at the 3′ end with fluorescein and the opposite hybridization probe should be labeled at the 5′ end with LightCycler Red 640/705 or other compatible dyes.
6. The position of the hybridization probes should bind closer to the 3′ end than the 5′ end of the target sequence (see Note 14).

Table 1
Conditions for detection of different *Candida* spp. and *Aspergillus* spp. by using real-time PCR on LightCycler® 480

Step	Target (°C)	Hold (s)	Ramp-rate (°C/s)	Acquisition (per °C)	Acquisition mode
Denaturation (1 cycle)					
	95	600	4.40		None
Amplification (35 cycles), quantification					
	95	5	4.40		None
	62	15	2.20		Single
	72	10	4.40		None
Melting curve (1 cycle)					
	95	30	4.40		None
	40	30	2.20		None
	85		0.19	3	Continuous
Cooling					
	40		20		None

3.4. Real-Time PCR on the LightCycler® 480 System Using Hybridization Probes

1. Hybridization probes: Dilute each probe (1 nmol) to a final concentration of 3 μM (=3 pmol/μl). The fluorescent labeled hybridization probes should not be exposed to light for a long period of time. Hybridization probes should be stored at −20°C.
2. Primer solution: Dilute the sense primer to a final concentration of 5 μM and the anti-sense primer to a final concentration of 20 μM (see Note 15). For this, prepare a fresh primer solution containing both primers before real-time PCR and add 2.5 μl of the sense primer (100 pmol/μl) and 10 μl of the anti-sense primer (100 pmol/μl) to 37.5 μl water.
3. Add 10 μl LightCycler 480 Probes Master (2×), 5 μl water, 1 μl primer solution (see step 2), 1.5 μl of FL1 probe or FL2 probe (3 μM), 1.5 μl LC1 probe or LC2 probe (3 μM), and 1 μl DNA template in a 96-well plate (see Note 16). It is recommended to prepare a mastermix (see Note 17).
4. Seal plates with a foil plate seal and centrifuge at 1,500 × *g* for 2 min. Perform real-time PCR on LightCycler® 480 system for the amplification of fungal DNA under the following conditions (Table 1). Ensure that the channels Red640 and Red705 are selected prior to starting the run.

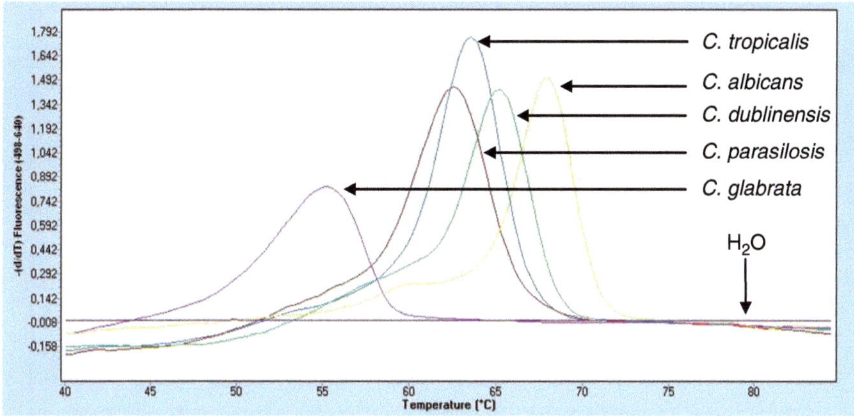

Fig. 2. Analysis of melting peaks of different *Candida* spp. after real-time PCR by using probes FL1 and LC1. Water control shows no melting peak.

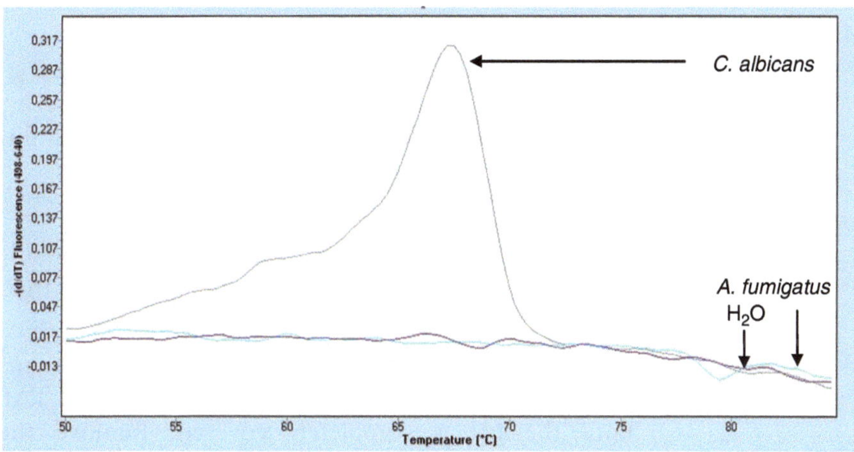

Fig. 3. Analysis of melting peak of *Candida albicans* after real-time PCR by using probes FL1 and LC1. *Aspergillus fumigatus* could not be detected by using these probes.

5. After real-time PCR using the hybridization probes FL1 and LC1, different *Candida* spp. can be detected by melting curve analysis (Fig. 2) (see Note 18). By using the hybridization probes FL2 and LC2 the discrimination between *Candida* and *Aspergillus* within the same sample is possible (Figs. 3 and 4).

3.5. Gel Electrophoresis

Gel electrophoresis of the PCR products is recommended to prove that the correct size of the DNA fragment has been amplified after real-time PCR.

1. For a 1× concentrated working solution of Tris–Acetate–EDTA buffer (TAE) dissolve a 50× concentrated stock solution of TAE buffer in a dilution of 1:50 with water. Store at room temperature.

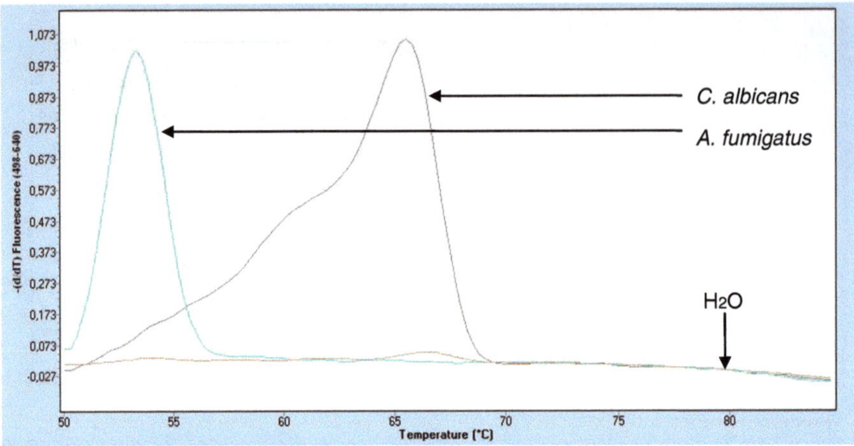

Fig. 4. Analysis of melting peaks for detection of *Candida albicans* and *Aspergillus fumigatus* within one sample after real-time PCR by using probes FL2 and LC2. *C. albicans* (approx. 66°C) and *A. fumigates* (approx. 53°C) showed different melting points. Water control shows no melting peak.

2. Prepare a 2% agarose gel (see Note 19) by mixing 2 g agarose with 100 ml 1× TAE buffer. Dissolve the agarose by boiling the mixture in a microwave at 800 W for 3 min until the solution is clear.
3. Let the gel solution cool down to 60°C.
4. Put the comb into the gel chamber. Fill in the liquid agarose solution and gently remove air bubbles. Be sure that the gel chamber is sealed to avoid spilling the liquid agarose.
5. After polymerization for 15 min, transfer the gel to an electrophoresis chamber and fill the chamber with 1× TAE buffer until the gel is completely covered. Remove the comb carefully.
6. Add 10 μl of the DNA size marker in one slot.
7. Mix 3 μl loading buffer with 10 μl cDNA per sample amplified by the PCR and transfer the samples carefully into the slots.
8. Connect the electrophoresis chamber with the power station and set the voltage and runtime. For a 2% agarose gel (100 ml) the settings are approximately 110 V for 60 min (see Note 20).

3.5.1. Staining of DNA Fragments in the Agarose Gel

1. To stain the DNA in the agarose gel, transfer the gel to an ethidium bromide bath after completion of the electrophoresis and incubate it for about 30 min (see Notes 3 and 21). For this, dissolve 300 μl of the ethidium bromide solution (10 mg/ml) in 500 ml water. Alternatively ethidium bromide or other dye can be added to the gel at the set-up stage
2. After incubation the gel should be washed for 10–20 min in water to remove excess ethidium bromide.

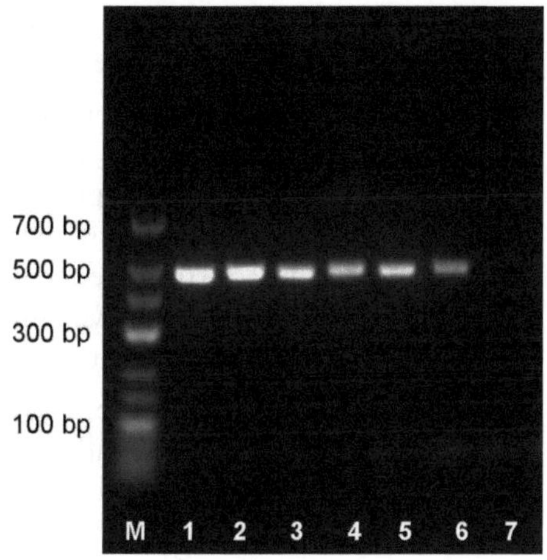

Fig. 5. Analysis of the length of the amplified 18sRNA fragments of *Candida* spp. and *Aspergillus* spp. after real-time PCR by gel electrophoresis: (1) *C. tropicalis*, (2) *C. parapsilosis*, (3) *C. dublinensis*, (4) *C. glabrata*, (5) *C. ablicans*, (6) *A. fumigatus*, (7) negative control (water). The base pair marker (GeneRuler™ Low Range DNA Ladder, Fermentas) is shown in the first lane.

3. Transfer the gel to the image acquisition system, which can detect dyes activated by UV rays. Take a picture according to the manufacturer's recommendations (Fig. 5).
4. Compare the length of the PCR fragments with the size marker and check the correct length of the amplified fragment has been generated.

4. Notes

1. For the isolation of DNA, kits and reagents from different companies are available. DNA isolation procedure often depends on the kind of fungi and sample material. Special modifications in the DNA isolation protocol for *Aspergillus* spp. should be made (7). After isolation, the DNA concentration should be measured before the PCR run (e.g., by photometry).
2. The use of the marker depends on the length of the amplified PCR fragment. Commonly used markers are 100 bp markers but should be adapted to the length of the amplified PCR product.
3. Ethidium bromide is a mutagenic dye, which intercalates between double-stranded DNA. It is strongly recommended to wear specialized gloves (e.g., nitrile gloves) for protection if you are working with ethidium bromide.

4. The target sequence should be unique for the species investigated by real-time PCR.
5. Longer products may decrease the exponential phase during the amplification process and the efficiency of PCR assays. The efficiency depends also on the nucleotide concentration. Longer fragments decrease the concentration of the nucleotides during the process and therefore decrease the PCR efficiency. However, amplification of longer fragments can work and is target dependent.
6. The length of the fragment has to be detected by gel electrophoresis after real-time PCR to prove that the correct size product was amplified.
7. The melting temperature (T_m) is calculated from the length of the primer and depends on the base pairs A, T, C, and G. There are other methods to calculate the melting temperature (e.g., Wallace rule). The melting temperature describes the temperature where 50% of the primers are bound to the DNA and is critical in the determination of the annealing temperature. The annealing temperature of the primers should be 3–5°C under the T_m and is defined as the maximal temperature where the primers bind to the complementary DNA. No samples will be amplified if the annealing temperature is too high. In contrast, low annealing temperatures lead to an unspecific binding of primers.
8. Primers often form dimers or crossdimers because of their short length and complementarity. Primer dimers are mostly seen in water controls. They align with themselves (dimers) or sense and anti-sense primer align with each other (crossdimers). They give false-positive fluorescence signals during the amplification process. After analysis of the melting curves, they can be seen as additional peaks. Primer dimers mostly show a lower melting temperature than the target gene fragments.
9. Common fluorescence dyes like SYBR green are intercalating dyes and are incorporated in double-stranded DNA. Therefore, fluorescence signals of primer dimers or other unspecific amplified products can be measured and can decrease the specificity of the PCR assay.
10. TaqMan probes consist of only one probe with bound dyes (quencher and reporter). The quencher suppresses the fluorescence signal of the reporter as long as the probe binds to the target sequence. The Taq polymerase cleaves the probe during the elongation process and quencher and reporter dye are removed from each other leading to an increased fluorescence signal. TaqMan probes are highly specific and one disadvantage is the requirement of different probes for differ-

ent sequences. In comparison, hybridization probes are able to distinguish between different species dependent on changes in the base pair composition of the target sequence.

11. There will only be a fluorescence signal, if the hybridization probes bind simultaneously on the target gene fragment and are near enough to each other.
12. After denaturation of DNA, primers and hybridization probes bind to their complementary target sequence on the DNA. A higher annealing temperature leads to a stronger hybridization to the target sequence. Primers and hybridization probes bind to the same target sequence and can compete with each other. Thus, the melting temperature of the hybridization probes should be higher than the melting temperature of the primers, which results in a stronger binding of the hybridization probes leading to higher fluorescence intensity.
13. Both hybridization probes bind to the same single strand, whereas one of the two primers binds to the sense strand ($5' \rightarrow 3'$ direction, sense primer). The second primer binds to the anti-sense strand ($3' \rightarrow 5'$ direction, anti-sense primer). After binding of primers and hybridization probes, the Taq polymerase extends the strand in $5' \rightarrow 3'$ direction (Amplification). The optimum temperature for the Taq polymerase is 72°C. Therefore, melting temperatures should be chosen lower than 80°C. Too high melting temperatures could lead to an interference of the amplification process with the annealing process of the hybridization probes.
14. Because of the existing competition between primers and hybridization probes, the hybridization probes should lay closer to the 3' end of the target sequence and thus far away from the sense primer. During the amplification process, the hybridization probes dissociate faster from the single strand when they are closer to the 5'-end of the sense primer because of interference with the Taq polymerase. This could lead to incorrect measurements of fluorescence signals. It is recommended to order and produce the designed probes by specialized firms (e.g., TIB MOLBIOL, Berlin, Germany).
15. This is known as asymmetric PCR and it results in the increased amplification of the strand to which the probes bind, ultimately increasing the signal. The optimum concentration must be tested for each primer.
16. Mostly in real-time PCR, the final concentration of cDNA should not exceed 100 ng.
17. Prepare two mastermix for each probe pair (do not put them together in one mix) and prepare each mastermix for two samples more.

18. The melting points are dependent on the sequence homology between the amplified PCR product and the hyb-probes.
19. The percentage of the agarose gel depends on the length of the amplified target fragment. The smaller the fragment, the higher should be the agarose concentration.
20. The running time and voltage for gel electrophoresis depends on the size and concentration of the agarose gel.
21. Ethidium bromide can also be added directly in agarose gel before electrophoresis. Ethidium bromide is positively charged thus it runs in the opposite direction to the negatively charged DNA.

References

1. Pappas PG (2010) Opportunistic fungi: a view to the future. Am J Med Sci 340(3): 253–257
2. Person AK, Kontoyiannis DP, Alexander BD (2011) Fungal infections in transplant and oncology patients. Hematol Oncol Clin North Am 25(1):193–213
3. Barnes PD, Marr KA (2007) Risks, diagnosis and outcomes of invasive fungal infections in haematopoietic stem cell transplant recipients. Br J Haematol 139(4):519–531
4. DeGraves FJ, Gao DY, Kaltenboeck B (2003) High-sensitivity quantitative PCR platform. Biotechniques 34(1):106–110
5. Bustin SA, Benes V, Nolan T et al (2005) Quantitative real-time RT-PCR—a perspective. J Mol Endocrinol 34(3):597–601
6. Fricke S, Fricke C, Schimmelpfennig C et al (2010) A real-time PCR assay for the differentiation of Candida species. J Appl Microbiol 109(4):1150–1158
7. Loeffler J, Henke N, Hebart H et al (2000) Quantification of fungal DNA by using fluorescence resonance energy transfer and the Light Cycler system. J Clin Microbiol 38(2): 586–590
8. Thornton B (2011) Real-time PCR (qPCR) primer design using free online software. Biochem Mol Biol Educ 39(2):145–154

Chapter 8

Quantitative and Multiplex Detection of Pathogenic Fungi Using Padlock Probes, Generic qPCR, and Suspension Array Readout

Magnus Jobs, Ronnie Eriksson, and Jonas Blomberg

Abstract

The multiplexing qualities of padlock probes and Luminex™ technology combined with the well-established quantitative feature of qPCR were the base for a ten-plex fungal detection protocol that quantitatively reveals ten different fungal species in a single experiment. Padlock probes are oligonucleotides designed to form circular DNA when hybridizing to specific target DNA. The 5′ and 3′ regions of the probes meet and ligate only when a specific target sequence is present in the examined sample. The region of the padlock probes that separates the target-specific 5′ and 3′ ends contains general primer sequences for amplification of circularized probes by means of rolling circle amplification (RCA) and qPCR. The interspersed region also contains specific tag sequences for subsequent Luminex™ recognition.

Key words: Multiplex, Padlock probes, Rolling circle amplification, Fungal detection, Suspension microarray

1. Introduction

A protocol for simultaneous quantitative detection of ten different fungal species is described (1). The protocol involves padlock probes, qPCR, and Luminex™ technology. A padlock probe is a long oligonucleotide designed to hybridize to a specific target sequence so that the 5′ and 3′ ends of the probe meet. The nick between the two ends can be closed via enzymatic ligation, resulting in a circularized probe (2). Circularized probes can be detected via rolling circle amplification (RCA) and/or PCR followed by amplification product detection (3). The amplification of circularized probes is made possible by including targets for amplification

Louise O'Connor and Barry Glynn (eds.), *Fungal Diagnostics: Methods and Protocols*, Methods in Molecular Biology, vol. 968, DOI 10.1007/978-1-62703-257-5_8, © Springer Science+Business Media New York 2013

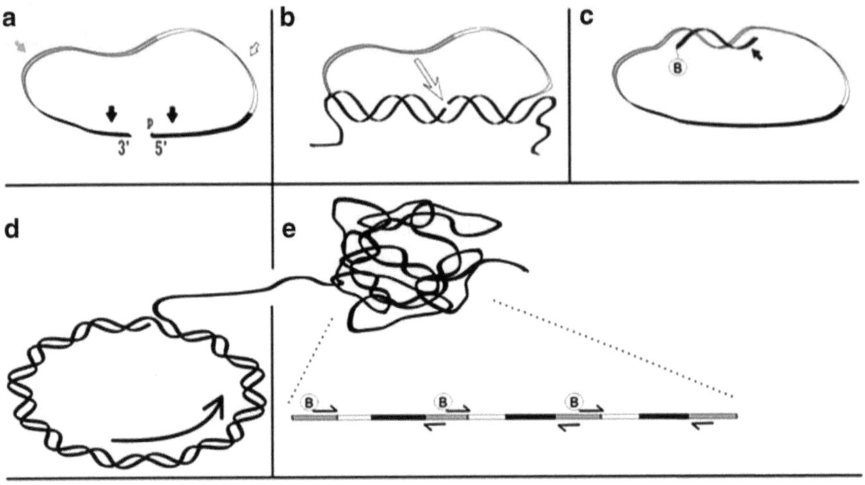

Fig.1. Outline of the padlock probe concept. (**a**) The 3′ and 5′ regions of the padlock probe (indicated in *black* and with *black arrows*) are complementary to a specific target sequence. The internal region contains a sequence designed for a general primer pair (indicated in *light gray* and with a light *gray arrow*) and an address sequence (indicated in *white* and with a *white arrow*). (**b**) When the target-specific regions hybridize to the target the 3′ and 5′ ends meet (the nick is indicated with an *arrow*), and via a ligation reaction the padlock probe becomes circularized. (**c**) A biotinylated primer (*dark gray* and pointed out with a *dark gray arrow*), complementary to one part of the general primer pair region, is then hybridized. (**d**) A rolling circle amplification (RCA) is performed generating a long repetitive sequence. (**e**) Finally, using the general primer pair (one of the primers is the same as used for the RCA), PCR is performed on the RCA product. The PCR product contains the address sequence and a biotin moiety for subsequent detection in the Luminex™ instrument.

primers in the padlock probe region separating the 5′ and 3′ regions. Detection of amplification products can be done in several ways. One clever way is to introduce a specific address tag sequence in the interspersed region that amplifies along with the rest of the circularized probe (4). If one or two fluorescent primers are used, amplification products can then hybridize to solid phase bound anti-tag sequences (i.e., microarray spots or Luminex ™ beads) and be detectable (5).

Padlock probes have successfully been used in various multiplex detection systems. The advantage of padlock probes in multiplexing is both the specificity of the ligation reaction that forms the circularized probes and the possibility of using a single PCR amplification primer pair common for all padlock probes included in the multiplex assay when amplifying the ligation products (6–8). The protocol described here is for a ten-plex fungal detection panel that quantitatively detects ten clinically important fungi. The technique is built on the padlock probe concept but two readouts are used: (1) a SybrGreen™ real-time PCR for quantification of circularized padlock probes and (2) suspension array technology (i.e., Luminex ™) for identification of amplified sequences (1). In Fig.1a, b, the padlock probe concept has been outlined, illustrating the different sections of the padlock probe and how the probe can hybridize and become circularized when a target sequence is present.

The RCA and subsequent real-time PCR (qPCR) using a single universal primer pair (one labeled primer) is also illustrated in Fig. 1c–e. The Luminex™ suspension array consists of fluorescence coded microspheres with coupled oligonucleotide anti-tag sequences for capturing of amplicons containing the sequence tag. Thus, a bead is analogous to a microarray spot (5). The combination of two readouts of padlock probe ligation, first via SybrGreen™ qPCR and then by suspension microarray technology, allows the well-established quantitative aspects of qPCR with a wide dynamic range to be combined with specific identification via the high multiplex capacity of padlock probes. Included in the protocol is a strategy for improved hybridization of PCR amplicons to the bead bound anti-tag. By letting labeled oligonucleotides hybridize to the 3′ and 5′ end region of the amplicon these regions become blocked for rehybridization to the complementary strand. This reduces competition between the anti-tag and the competing strand. Moreover, the fluorophores on the end covering oligonucleotides contributes to an improved detection signal.

2. Materials

Solutions should be prepared using ultrapure water. Pre-PCR reagents should be handled in a clean room well separated from post-PCR activities. Use "PCR-grade water" when diluting stocks and working solutions of oligonucleotides. Shelf products in use in this protocol should be handled and stored according to the manufacturer's instructions. Oligonucleotides and PCR reagents should be stored in −20°C and buffers and microspheres at 4°C.

2.1. DNA Isolation Components

1. NucliSENS® Extraction system (BioMerieux).
2. 0.5 mm-diameter zirconium-silica beads.
3. Tissuelyser (Qiagen AB, Stockholm, Sweden).
4. 2 mL Eppendorf safe-lock tubes.
5. NucliSENS® easyMAG™ lysis buffer.
6. 2× Sputolysin® (Calbiochem, San Diego, CA, USA).

2.2. Synthetic Targets, Primers, and Probes

1. Oligonucleotides (Table 1) (see Note 1).
2. Denhartd's solution (Sigma Aldrich, St Louis, MO, USA).

2.3. Padlock Probes, Ligation, and RCA Components

1. Padlock probes for the fungal species included in the assay, all 5′ phosphorylated (Table 1).
2. General forward primer Biot-5′-AAGATATCGTAAGGAT-3′ (5′ biotinylated) 1 μM.

Table 1
Padlock probe- and synthetic target sequences

<table>
<tr><th>Fungal species
↑ High eff. probe
↓ Low eff. probe</th><th>*Padlock probe sequence (5′-3′)
**Synthetic target sequence (5′-3′)</th></tr>
<tr><td>A. flavus
↑</td><td>*PO_4-TTGCGTTCGGCAAGCGCC**ATTGGTAAATTGGTAAATGAATTG**ATCCTTACGATA
TCTTGGATAAGTGGGATATCCAAGGTCAACCTGGAAAAAGATTGAT
**TCCGGCATCGATGAAGAACGCAGCCCCGGCCGGCGCTTGCCGAACGCAA|
ATCAATCTTTTTCCAGGTTGACCTTGGATCAGGTAGCATATCAATAAGCGGAGGA</td></tr>
<tr><td>A. fumigatus
↑</td><td>*PO_4-GGGTGTCGGCTGGCGC**TTAGATGAATTGTGAAGTATTTAG**ATCCTTACGATATCT
TGGATAAGTGGGATAATCCGAGGTCAACCTTAGAAAAATAAAGTT
**TCCGGCATCGATGAAGAACGCAGCCCCGGCCGGCGCCAGCCGACACCC|
AACTTTATTTTTCTAAGGTTGACCTCGGATCAGGTAGCATATCAATAAGCGGAGGA</td></tr>
<tr><td>A. nidulans
↓</td><td>*PO_4-TCGAGCGGGTGACAAAGCCC**TGAAATGAATGAATGATGAAATTG**ATCCTTACGA
TATCTTGGATAAGTGGGATAGCCCGGCCGGCCCTAA
**TCCGGCATCGATGAAGAACGCAGCCGAGCGTATGGGGCTTTGTCACCCGCTCGA|
TTAGGGCCGGCCGGGCGCCAGCCGACGCATATCAATAAGCGGAGGA</td></tr>
<tr><td>A. niger
↑</td><td>*PO_4-AGGCGCCGGCCAATCCTAC**GTAAAAAGAAAGGTATAAAGGTAA**ATCCTTACGAT
ATCTTGGATAAGTGGGATAGAAAGAATGGTTGGAAAACGTCGGC
**TCCGGCATCGATGAAGAACGCAGCTGCTCTGTAGGATTGGCCGGCGCCT|
GCCGACGTTTTCCAACCATTCTTTCCAGGTTGCATATCAATAAGCGGAGGA</td></tr>
<tr><td>C. albicans
↑</td><td>*PO_4-CGCTACCGCCGCAAGCAAT**GATTTGAAGATTATTGGTAATGTA**ATCCTTACGAT
ATCTTGGATAAGTGGGATAAGGTCAAAGTTTGAAGATATACGTGGTAGA
**TCTCGCATCGATGAAGAACGCAGCAACATTGCTTGCGGCGGTAGCG|
TCTACCACGTATATCTTCAAACTTTGACCTAAGCATATCAATAAGCGGAGGA</td></tr>
<tr><td>C. glabrata
↑</td><td>*PO_4-GTTGGTAAAACCTAATACAGTATTAACCCCGATTGATTATTGTGATTTGAATTG
ATCCTTACGATATCTTGGATAAGTGGGATACTTATCCCTCCCTAGATCAACACCGA
**TCTCGCATCGATGAAGAACGCAGCCGGGGGTTAATACTGTATTAGGTTTTACCAAC|
TCGGTGTTGATCTAGGGAGGGATAAGTGGCATATCAATAAGCGGAGA</td></tr>
</table>

C. tropicalis ↓	*PO_4-CCACTAGCAAAATAAGCGTTTTTGGATAAA**TGATATGAATTGGATTATTGGTAT** ATCCTTACGATATCTTGGATAAGTGGGATAAGGTCAAAGTTATGAAATAAATTGTGGTGG **TCTCGCATCGATGAAGAACGCAGCTTTATCCAAAAACGCTTATTTTGCTAGTGG| CCACCACAATTTATTTCATAACTTTGACCTGCATATCAATAAGCGGAGGA
C. parapsilosis ↓	*PO_4-GGAGTTTGTACCAATGAGTGGAAAAAAAC**GTTAGTTAGATTATTGTTAGTTAG**A TCCTTACGATATCTTGGATAAGTGGGATATGATTTGAGGTCGAATTTGGAAGAAGTTTT **TCTCGCATCGATGAAGAACGCAGCGTTTTTTTCCACTCATTGGTACAAACTCC| AAAACTTCTTCCAAATTCGACCTCAAATCAGCATATCAATAAGCGGAGA
Cr. Neoformans ↓	*PO_4-GCCGAAGACTACCCCATAGGCC**GTAAGATGTTGATATAGAAGATTA**ATCCTTA CGATATCTTGGATAAGTGGGATAAAACAAAAAAGAGATGGTTGTTATCAGCAA **TTCCACATCGATGAAGAACGCAGCTGGGCCTATGGGGTAGTCTTCGGC| TTGCTGATAACAACCATCTCTTTTTTGTTTGAGCATATCAATAAGCGGAGGA
P. Jiroveci ↓	*PO_4-GAATTTCAGACTAGCATGCATATAATTATT**TAATGTTGTGAATAATGTAGAAAG** ATCCTTACGATATCTTGGATAAGTGGGATAGACACTAGGCAAAGAAAAAAAGCTACTTTT **TCTCGCGTCGATGAAGAACGTGGCAATAATTATATGCATGCTAGTCTGAAATTC| AAAAGTAGCTTTTTTTCTTTGCCTAGTGTCGCATATCAATAAGCGGAGGA

Padlock probe sequences for the ten targeted fungi are indicated with an asterix (*). The target matching sections are underlined and the address tags are shown in bold. The remaining unmodified section is the gen`eral primer pair sequence, identical for all padlock probes. The synthetic target sequences are marked with a double asterix (**). Here the underlined section represents the padlock probe matching section. The padlock probe ligation position has been pointed out with a bar (|). Extra sequences flanking the target region has been included and represents interspecies conserved sequences (in vivo these conserved regions may not be immediately adjacent to the padlock probe matching sections)

3. 10× Ampligase® reaction buffer and Ampligase® thermostable DNA ligase (Epicentre Biotechnologies, WI, USA).
4. PCR-grade water (Applied Biosystems, Stockholm, Sweden).
5. Padlock probe mix with 10 nM of each padlock probe.
6. 10× Phi29 buffer (Fermentas, Vilnius, Lithuania) and Phi29 DNA polymerase 10 U/μL.
7. dNTP 10 mM.
8. BSA 2 μg/μL.

2.4. PCR Components

1. 10× SYBR® Green PCR Buffer.
2. dNTPs including dUTP 12.5 mM.
3. AmpliTaq Gold® DNA Polymerase 5 U/μL.
4. AmpErase® UNG 1 U/μL.
5. Rotor-Gene 3000 (Corbett Life Science, Concorde, New South Wales, Australia).
6. 0.2 mL PCR tubes.
7. Rotor-gene 6.1.
8. General forward primer (see Subheading 2.3) and general reverse primer 5′-TTGGATAAGTGGGATA-3′ 10 μM of each.

2.5. Suspension Array Components (Coupling of Anti-tag Oligonucleotides)

1. 5′ C-12 amino-modified oligonucleotide anti-tags 100 μM (Table 2).
2. Ten sets (Table 2) of carboxylated polystyrene FlexMap™ microspheres (Luminex Corporation, Austin, TX, USA).
3. 0.1 M 2-Morpholinoethane sulfonic acid (MES, pH 4.5) MES buffer.
4. Desiccated 1-Ethyl-3-3-dimethylaminopropyl carbodiimide (EDC) (Pierce Thermo Fisher Scientific).
5. 0.02% Tween-20.
6. 0.1% SDS, Tris–EDTA (TE, pH 8.0) buffer.
7. Table-top ultrasonic cleaner.

2.6. Components for Suspension Array Hybridization and Detection

1. Tetramethyl ammonium chloride (TMAC) hybridization buffer (4.5 M TMAC, 0.15% sodium lauryl sarcosinate; "sarkosyl" (Sigma), 75 mM Tris–HCl pH 8, and 6 mM EDTA pH 8).
2. Tris–EDTA (pH 8.0) buffer.
3. 5′ Cy3-labeled end covering oligonucleotides (Cy3-5′-ATCCTTACGATATCTT-3′ and Cy3-5′-TTGGATAAGTGGGATA-3′) at 10 μM.
4. Streptavidin-R-Phycoerytrin 0.15 μg/μL (Qiagen AB, Stockholm, Sweden).

Table 2
Outlines of the amino-modified carbon 12-linked anti-tag sequences that are coupled to different color-coded microspheres

Anti-tag for	Anti-tag sequence	Microplex™ xTAG
A. flavus	N-C12-ATTGGTAAATTGGTAAATGAATTG	LUA-7
A. fumigatus	N-C12-TTAGATGAATTGTGAAGTATTTAG	LUA-90
A. nidulans	N-C12-TGAAATGAATGAATGATGAAATTG	LUA-35
A. niger	N-C12-GTAAAAAGAAAGGTATAAAGGTAA	LUA-30
C. albicans	N-C12-GATTTGAAGATTATTGGTAATGTA	LUA-4
C. glabrata	N-C12-GATTGATTATTGTGATTTGAATTG	LUA-5
C. tropicalis	N-C12-TGATATGAATTGGATTATTGGTAT	LUA-70
C. parapsilosis	N-C12-GTTAGTTAGATTATTGTTAGTTAG	LUA-80
Cr. Neoformans	N-C12-GTAAGATGTTGATATAGAAGATTA	LUA-9
P. jiroveci	N-C12-TAATGTTGTGAATAATGTAGAAAG	LUA-40

The anti-tags match tags included in padlock probes targeting the different fungal species indicated. Pre-coupled microspheres, with the same anti-tags, can be ordered directly from the Luminex™ and the table points out the corresponding Microplex™ microspheres

5. Non-skirted thin-wall 96× 0.2 mL microplates (Bioplastics, Landgraaf, The Nederlands).
6. Heater block part # 67-50066-00-001 (Luminex Corporation, Austin, TX, USA).
7. Thermostar microplate shaking table (BMG lab thechnologies, Germany).
8. The Luminex™ 200 system (Luminex Corporation, Austin, TX, USA).
9. STarStation software (Applied Cytometry System, Sheffield, UK).

3. Methods

3.1. Preparation of Standards Using Synthetic Targets

1. Prepare stock solutions of synthetic targets 100 μM (100 pmol/μL) according to instructions from the oligonucleotide manufacturer (see Note 1). Prepare two pools of synthetic targets, one pool of targets for low-efficiency padlock probes (marked with down arrow in Table 1) and one pool of targets for high-efficiency padlock probes (marked with up arrow in Table 1). Prepare the two pools so that each of the

five targets included in a pool makes up one-fifth of the total copy number in the pool.

2. Do this by mixing 10 μL from each stock of synthetic targets belonging to the same category. Add 251 μL PCR-grade water to the 50 μL pool stocks so that the pools contain 10^{13} molecules/μL (all five molecule species combined).
3. From here, use 0.05× Denhardt's solution instead of pure water when preparing the standard dilution series.
4. Dilute 1,000 times (1 μL target pool plus 999 μL 0.05× Denhardt's solution).
5. From that solution dilute five times (50 μL target pool plus 200 μL 0.05 Denhardt's solution).
6. From this solution serially dilute by a factor of 10 (10 μL target pool plus 90 μL 0.05× Denhardt's solution) until a series ranging from 0.2×10^2 molecules/μL to 0.2×10^7 molecules/μL has been established. Make sure to vortex every new dilution thoroughly before preparing the next dilution in the series.

3.2. Dilution of Oligonucleotides (Probes and Primers)

1. Dilute padlock probes and primers in PCR-grade water to 100 μM (100 pmol/μL) stock solutions (follow manufacturers instruction for reconstitution).
2. Prepare 1 μM work solutions of each padlock probe by diluting the main stock solutions a hundred times.
3. Prepare a mixture of all ten padlock probes with a 10 nM concentration of each padlock probe (for a 500 μL mixture take 5 μL of each padlock probe work solution and add 450 μL PCR-grade water).
4. Prepare two different general forward primer work solutions, one 1 μM solution for use when setting up the RCA reaction and one 10 μM solution for use when setting up the PCR. Dilute the general reverse primer to a work solution of 10 μM.
5. All dilutions can be stored in −20°C.

3.3. Coupling of Anti-tag Oligonucleotides to Microspheres

1. Thaw two 10 mg aliquots of desiccated EDC powder.
2. Remove the ten different FlexMap™ microspheres from the fridge. Let the reagents adjust to room temperature (see Note 2).
3. Resuspend the microspheres by thorough vortexing and then sonication for 20 s.
4. Transfer 2.5 million (200 μL from stock) of each set of microspheres to ten pre-labeled 1.5 mL Eppendof tubes (see Note 3).
5. Pellet the microspheres by microcentrifugation at ≥8,000 × *g* for 1–2 min.

6. Remove the supernatants and resuspend the microspheres in 25 μL MES buffer (vortex and sonicate for 20 s).
7. Add 2 μL of the ten anti-tag oligonucleotides to the corresponding microsphere sets and vortex.
8. Prepare a fresh solution of 10 mg/ml EDC by adding 1 mL water to one of the aliquots of desiccated EDC.
9. Quickly add 2.5 μL of the fresh EDC solution to all ten microsphere sets and vortex.
10. Incubate at room temperature for 30 min in dark.
11. Prepare another fresh solution of EDC by adding 1 mL water to the second EDC aliquot.
12. Again quickly add 2.5 μL of the fresh EDC solution to all ten microsphere sets and vortex.
13. Incubate for another 30 min at room temperature in the dark.
14. Add 0.5 mL 0.02% Tween-20 to each set.
15. Pellet the sets by microcentrifugation at ≥8,000 × *g* for 1–2 min.
16. Remove the supernatant and resuspend the sets in 0.5 mL 1% SDS and vortex.
17. Pellet the sets by microcentrifugation at ≥8,000 × *g* for 1–2 min.
18. Remove the supernatant and resuspend the sets in 50 μL TE buffer (pH 8.0).
19. Prepare a mixture of all ten sets by mixing equal volumes of each.
20. Store the sets and the mixture of coupled microspheres (50,000 spheres/μL) in the fridge protected from light. (The coupling procedure is a modification of an original Luminex™ protocol) (see Notes 4 and 5).

3.4. DNA Isolation

Clinical samples may be in the form of charcoal swabs, urine, vaginal swab culture broth, bronchoalveolar lavage, or sputum. Procedures for initiating DNA isolation vary for the different types. Steps 1–3 represent the start of the DNA isolation procedure for the different types of clinical samples.

1. Wash charcoal swabs in 600 μL NucliSENS® easyMAG™ lysis buffer by inserting the swab in a 1.5 mL Eppendorf tube containing the buffer and rotate the swab thoroughly. Then proceed to step 4.
2. Transfer 1 mL urine, vaginal swab broth, or bronchoalveolar lavage to 1.5 mL Eppendorf tube and centrifuge the sample at 13,000 × *g* for 20 min. Discard the supernatant and dissolve

the pellet in 600 μL NucliSENS® easyMAG™ lysis buffer. Then proceed to step 4.

3. Dilute sputum samples in an equal volume of 2× Sputolysin® (1× Sputolysin® final concentration) and vortex thoroughly. Incubate for 20 min and then proceed to step 4.
4. Perform cell lysis by bead beating. First transfer the suspension to 2 mL Eppendorf safe-lock tubes containing 600 μL 0.5 mm-diameter zirconium-silica beads. Then shake the samples in a Tissuelyser for 10 min at 30 Hz.
5. Transfer the lysates to clean 15 mL Falcon tubes (leave the beads in the original tube). Rinse the beads twice in 1 mL NucliSENS® easyMAG™ lysis buffer and add the rinse solution to the collected lysate in the Falcon tube. The final volume of cell lysate will be approximately 2.5 mL (it is impossible to recover the entire volume).
6. Add the full sample volume to a sample vessel for NucliSENS® easyMAG™ extraction and run the extraction according to the manufacturer's instructions. Set the elution volume to 60 μL. Extracts can be stored at −20°C.

3.5. Ligation and RCA

The reaction mixtures of both the ligation reaction and the RCA reaction must be prepared in a clean room well separated from samples, synthetic targets, and PCR products from previous experiments. The following protocol is based on a single reaction and needs to be multiplied to match the number of samples analyzed. In fact, preparing a reaction mixture for a single reaction is not recommended since pipetting such small volumes accurately is difficult. Including the two standard dilution series, counting five concentrations each (described above), means that at least a volume for 10 reactions must be prepared only to cover the dilution series.

1. Thaw the material for the ligation reaction (described above).
2. Prepare the ligation reaction mixture by mixing 1 μL of 10× Ampligase® reaction buffer, 0.2 μL Ampligase® thermostable DNA ligase, 1 μL Padlock probe mixture, and 4.8 μL PCR-grade water.
3. Add 3 μL of DNA extract to the reaction mixture, pipette-mix, and place the reaction on a thermal cycler at 95°C for 2 min, 55°C 30 min, and keep at 4°C until the next step.
4. Thaw the RCA components (described above).
5. Prepare the RCA reaction by mixing 2 μL 10× Phi29 reaction buffer, 2 μL BSA, 0.5 μL forward primer, 0.25 μL dNTP, and 0.3 μL Phi29 DNA polymerase.
6. Add 4.95 μL PCR-grade water (to reach a volume of 10 μL).

7. Finally mix the RCA reaction mixture with the fully incubated ligation reaction (20 μL final volume) and incubate at 37°C for 30 min followed by 85°C for 4 min and keep at 4°C until the next step.

3.6. PCR

The PCR mixture must be prepared in a clean room well separated from samples and PCR products from previous experiments.

1. Prepare a 25 μL PCR mixture by mixing 2.5 μL 10×SYBR® Green PCR Buffer, 0.4 μL dNTP (including dUTP), 1.5 μL $MgCl_2$, 0.25 μL AmpliTaq Gold® DNA Polymerase, 0.25 μL AmpErase® UNG, and 0.75 μL of each general padlock probe primer.
2. Add 16.1 μL PCR-grade water.
3. Add 2.5 μL of the RCA product.
4. Program your real-time PCR instrument (we use a Rotor-Gene 3000 from Corbett Life Science) as follows: 50°C for 2 min, 95°C for 10 min, followed by 30 cycles of 95°C for 15 s, 52°C for 15 s, and 68°C for 30 s. Monitor fluorescence during the 68°C step in each cycle. Monitor fluorescence in the FAM/Sybr channel (source: 470 nm, detector: 510 nm).
5. The real-time PCR will reveal any ligation event in the previous steps in the procedure so at this point it is appropriate to review the real-time PCR data to decide whether to proceed with the protocol or not.

3.7. Measurements in the Luminex™ 200 Instrument

1. Bring all the Luminex™ components (see Subheading 2.6) to room temperature.
2. Prepare the hybridization reaction by vortexing and sonicating (for 20 s) the anti-tag coupled microsphere mixture (containing all ten anti-tag coupled microspheres).
3. Dilute the microsphere mixture in TMAC hybridization solution by mixing 1.5 μL microsphere mixture with 33 μL TMAC (use wells from low non-skirted thin-wall 96-well plates).
4. Add 1 μL of each end covering oligonucleotide, 8.5 μL TE buffer, and finally 5 μL PCR product.
5. Pipette-mix the hybridization mixture and denature it for 2 min at 95°C (preferably in a 96-well PCR instrument).
6. Place the wells in a heater block on a shaking table heated to 50°C and allow hybridization for 30 min.
7. Add 2 μL streptavidin-R-Phycoerytrin, pipette-mix, and let the incubation continue for another 15 min.
8. Prepare the Luminex™ instrument by adjusting the analysis probe (the suction needle) to the current sample plate and run the start-up scripts.

9. Program the instrument to analyze median fluorescence intensity (MFI) from the ten microsphere regions to which anti-tags have been coupled.
10. Initiate the assay by setting the instrument to calculate MFI based on 100 measurements from each microsphere set and allow the Luminex™ XYP reach 50°C.
11. Place the heater block with the samples in the instrument and run.

3.8. Analysis

PCR amplification data are preferably reviewed in the dedicated instrument software.

1. Instruct the PCR software to automatically identify cycle threshold (Ct) values.
2. MFI from all ten microsphere regions are presented as numbers by the Luminex™ instrument and can easily be exported to an Excel file (or be reviewed directly in the instrument software). Figure 2a, b shows data from a titration series of a synthetic *Candida glabrata* sequence. The MFIs have been exported to Excel and converted into a diagram.
3. When running clinical samples, the use of standard titration series with known starting copy numbers run in parallel makes it possible to estimate sample copy numbers. In a ten-plex assay ten such dilution series of the different targets would be very impractical and expensive to include. Instead all ten targets could be pooled so that each target makes up one-tenth of the total copy number in the pool. The SybrGreen™-based qPCR would then detect the pool as a single target species and the copy number of the sample could be calculated based on the amplification of the pool (given that the sample contains only one fungal species). However, we have noticed that the padlock probes matching the different fungal species amplify with different efficiencies resulting in different Ct values despite the same target copy numbers. Therefore, we have categorized the padlock probes in two categories and made two pools of targets, one pool containing targets for the more efficient padlock probes and one for the less efficient. Both pools contain five target species. The total copy numbers of the five targets in a pool together makes up the standard for each dilution step. In this way two standard curves with slightly different slope will be produced. The two standard curves encompass most of the differences between padlock probes and make it possible to approximate sample copy numbers once the Luminex™ analysis has revealed what fungal species was present in the sample.

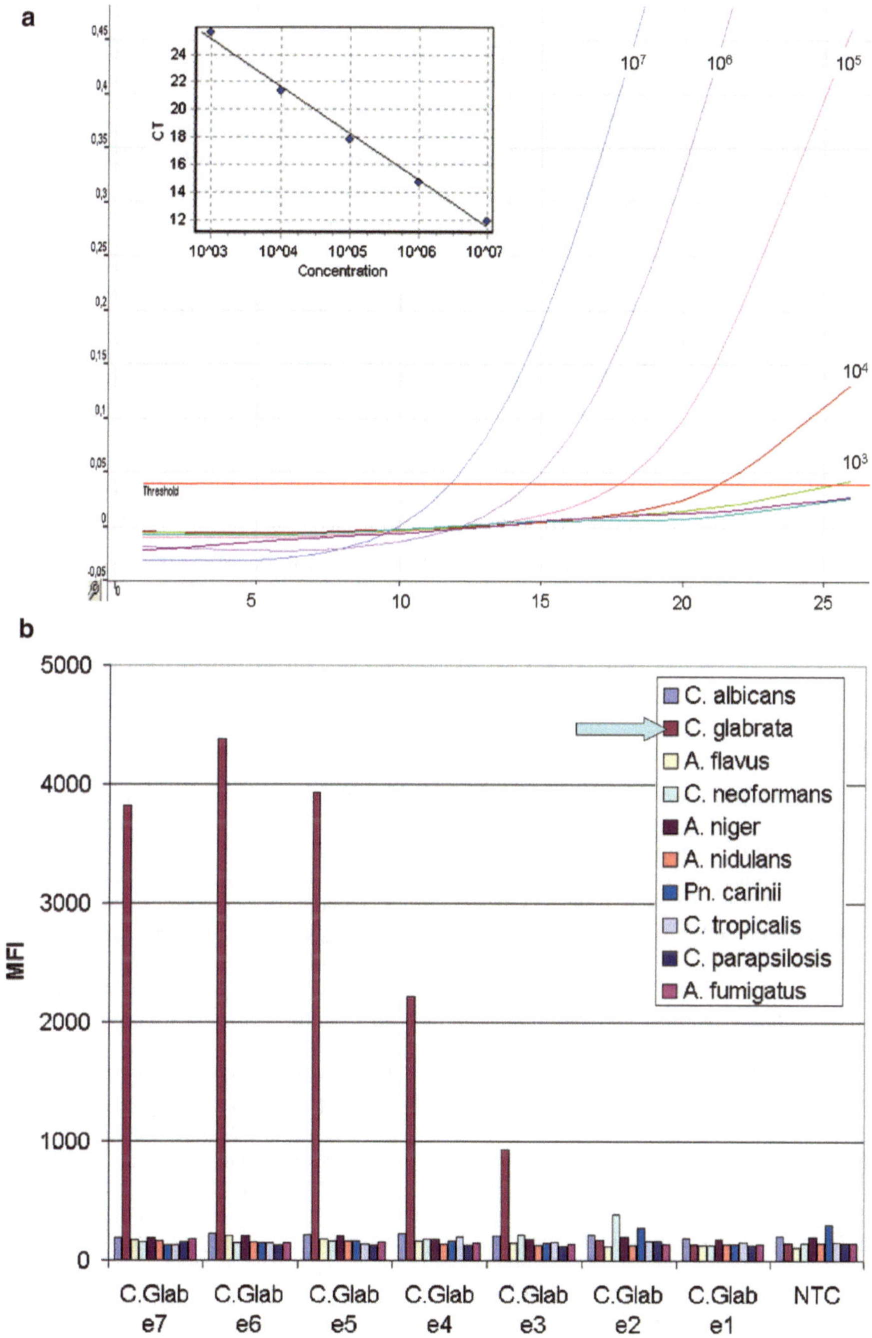

Fig. 2. The figure shows the result of padlock probe-based analyses of a titration series of a synthetic fungal DNA sequence (*Candida glabrata*). (**a**) Results from SybrGreen™-based real-time PCR data and (**b**) Luminex™ median fluorescence intensity (MFI) data. The figure demonstrates the methods' ability to quantify a sample in ten-plex mode.

4. Notes

1. Take special care to avoid contamination when handling concentrated synthetic targets. Prepare and store the high concentration targets in a room separated from where the main laboratory work is conducted.
2. Prepare several Eppendorf tubes (1.5 mL) with 10 mg EDC aliquots in each in advance and store them in a sealed container together with desiccant.
3. Label the tubes clearly with information on microsphere region and what fungus the corresponding anti-tag will detect.
4. Coupled microspheres for 500 reactions are produced by following this coupling protocol so it is not necessary to do this every time you run an assay.
5. Microspheres with pre-coupled anti-tags (microplex xTAG microspheres) can be ordered from Luminex avoiding the coupling procedure altogether. In Table 2 the corresponding microplex xTAG microspheres are indicated.

References

1. Eriksson R, Jobs M, Ekstrand C, Ullberg M, Herrmann B, Landegren U, Nilsson M, Blomberg J (2009) Multiplex and quantifiable detection of nucleic acid from pathogenic fungi using padlock probes, generic real time PCR and specific suspension array readout. J Microbiol Methods 78(2):195–202
2. Nilsson M, Malmgren H, Samiotaki M, Kwiatkowski M, Chowdhary BP, Landegren U (1994) Padlock probes: circularizing oligonucleotides for localized DNA detection. Science 265(5181):2085–2088
3. Baner J, Nilsson M, Mendel-Hartvig M, Landegren U (1998) Signal amplification of padlock probes by rolling circle replication. Nucleic Acids Res 26(22):5073–5078
4. Baner J, Gyarmati P, Yacoub A, Hakhverdyan M, Stenberg J, Ericsson O, Nilsson M, Landegren U, Belak S (2007) Microarray-based molecular detection of foot-and-mouth disease, vesicular stomatitis and swine vesicular disease viruses, using padlock probes. J Virol Methods 143(2): 200–206
5. Fulton RJ, McDade RL, Smith PL, Kienker LJ, Kettman JR Jr (1997) Advanced multiplexed analysis with the FlowMetrix system. Clin Chem 43(9):1749–1756
6. Nilsson M, Landegren U, Antson DO (2002) Single-nucleotide sequence discrimination in situ using padlock probes. Curr Protoc Hum Genet Chapter 4:Unit 4:11 page 1–12
7. Baner J, Isaksson A, Waldenstrom E, Jarvius J, Landegren U, Nilsson M (2003) Parallel gene analysis with allele-specific padlock probes and tag microarrays. Nucleic Acids Res 31(17):e103
8. Edwards KJ, Reid AL, Coghill JA, Berry ST, Barker GL (2009) Multiplex single nucleotide polymorphism (SNP)-based genotyping in allohexaploid wheat using padlock probes. Plant Biotechnol J 7(4):375–390

Chapter 9

Species-Specific Identification of a Wide Range of Clinically Relevant Fungal Pathogens by the Luminex® xMAP Technology

S. Preuner and T. Lion

Abstract

Invasive fungal infections (IFI) are a common cause of life-threatening events in immunocompromised patients. Early detection and identification of the fungal pathogen is an important prerequisite for timely onset of the most appropriate treatment. Methods based on fungal culture are often too slow to be clinically useful. Other approaches to the identification of fungal species, including molecular techniques, are often restricted to a small number of the most commonly occurring pathogens and are therefore of limited use in the clinical setting. The development of assays for the detection and identification of a broad-range of clinically relevant fungal species is therefore an urgently needed step towards optimized diagnostics of IFI.

The Luminex® xMAP technology offers a platform for the establishment of multiplex assays permitting high-throughput analysis of up to 100 different target molecules in a single test. Here we describe a Luminex®-based multiplex assay permitting rapid detection and identification of 10 fungal genera and 29 different species, including both commonly occurring and emerging fungal pathogens.

Key words: Species-specific fungus identification, Emerging fungi, Multiplex assay, Luminex® xMAP technology, Semi-nested PCR

1. Introduction

The majority of invasive fungal infections (IFI) are attributable to a limited number of *Aspergillus* and *Candida* species (1, 2), but the incidence of emerging pathogens including particularly yeast-like and filamentous fungi such as Zygomycetes has been increasing (2–6). Moreover, the drug resistance profiles of certain fungal species have changed over the last decades. Clinically important examples include *Aspergillus terreus*, known to display resistance to

Louise O'Connor and Barry Glynn (eds.), *Fungal Diagnostics: Methods and Protocols*, Methods in Molecular Biology, vol. 968, DOI 10.1007/978-1-62703-257-5_9, © Springer Science+Business Media New York 2013

Amphotericin B (4), *Candida krusei*, which is innately resistant to fluconazol, and *Candida glabrata*, which is commonly insensitive to other azoles (4, 7, 8). Wide prophylactic use of voriconazol, which is active against fluconazol-resistant *Candida* species, could in turn contribute to increased incidence of zygomycoses displaying inherent resistance to this agent (3–5). The occurrence of resistant fungal species, which might, at least in part, be attributable to the prophylactic use of broad-spectrum antifungals, renders the choice of the most appropriate treatment a challenging task. Despite the existence of resistant strains within individual fungal species, recognition of fungal pathogens at the species level mostly permits prediction of susceptibility to individual antifungal agents. The identification of fungal pathogens in patients at high risk of IFI is therefore an important prerequisite for appropriate selection of treatment.

To meet this requirement, a variety of molecular assays for the identification of fungal species have been developed. The region most frequently targeted by molecular tests is the ribosomal (r) DNA gene, which is present in multiple copies in the fungal genome, and therefore permits fungus detection even in the presence of very low pathogen concentrations.

The rDNA region contains interspersed sequences, the *i*nternal *t*ranscribed *s*pacer ITS1 and ITS2 regions, which are highly variable among fungal pathogens, and can thus be readily exploited for species identification. Methods for the identification of fungal pathogens based on targeting of the ITS1 and/or ITS2 regions include the amplification by PCR followed by direct sequencing (9, 10), fragment size analysis (11), pyrosequencing (12, 13), or microarray analysis (14, 15). Additionally, several hybridization-based platforms, combining PCR amplification of the ITS1/ITS2 target regions and subsequent hybridization to specific molecular probes, have been developed (16–18).

To establish an efficient approach to the identification of a wide range of fungal pathogens in the clinical setting, with the aim to also include resistant and emerging species, we have developed a broad-range multiplex assay based on the Luminex® xMAP technology. The assay facilitates the identification of 10 fungal genera and 29 different fungal species (19) covering the most important *Aspergillus* ($n=7$) and *Candida* ($n=9$) species as well as representatives of other emerging fungal genera, such as *Cryptococcus*, *Trichosporon*, *Fusarium*, *Penicillium*, *Acremonium*, *Mucor*, *Rhizopus*, and *Absidia*. The assay for identification of fungal species was established to supplement our patented panfungal real-time PCR screening assay which includes two separate reactions, one covering primarily moulds and the other yeasts and Zygomycetes (20). Since this assay provides general information on the type of fungal pathogen present, we have adjusted the Luminex® xMAP assay accordingly by dividing the functionalized beads into two multiplex sets, one for moulds and one for yeast/Zygomycetes in order to generate a more economic approach to the identification of fungal pathogens.

In the present version of the assay, carefully designed hybridization probes, including 69 species-specific and 9 genus-specific oligonucleotides targeting the highly variable ITS2 region (Table 1), were covalently attached to color-coded microbeads. The microbeads are uniquely defined via different shades of red and infrared and can be identified by their spectral address using flow cytometry. The available spectrum of unique color codes allows the simultaneous analysis of up to 100 different targets in one multiplex reaction. To permit the identification of fungal pathogens with high sensitivity, DNA purified from clinical specimens is used as template for amplification of the ITS2 region by semi-nested PCR. Subsequently, the amplicons are incubated with a set of microbeads carrying the indicated probes for specific hybridization. The pre-amplification step of the target DNA is essential because fungal pathogens may be present at very low amounts in clinical specimens such as peripheral blood. The PCR primers used for the second round of the semi-nested amplification are biotinylated (see Subheading 2.2), thus permitting the detection of fungal amplicons specifically hybridized to microbead-bound capture probes by a simple biotin-streptavidine/phycoerythrine reporter reaction leading to green fluorescent light emission (Fig. 1). Analysis via flow cytometry on a Luminex® instrument permits simultaneous detection of hybridized amplicons (green fluorescence) and the microbead involved (red fluorescence). Detection of both fluorescent signals permits the assignment of the hybridized amplicon to a specific capture probe, thus facilitating identification of the fungal pathogen. To account for potential mutations within the target region, the capture probes were designed in a redundant manner, including up to three different probes per fungal species or genus targeting adjacent sequences within the ITS2 region (Table 1). If a mutation prevented the hybridization of a fungal amplicon to one probe, the remaining probe(s) would still permit correct identification of the pathogen.

The multiplex assays were tested for sensitivity, specificity, and reproducibility by investigating 82 different fungal species and strains derived from reference repositories including the ATCC (American Tissue Culture Collection), DSM (Deutsche Sammlung von Mikroorganismen und Zellkulturen GmbH, Germany), and KAM (Klinische Abteilung für Mikrobiologie, Medical University of Vienna, Austria).

The sensitivity (detection limit) of the assays was determined by analyzing serial dilutions of fungal DNA as template for semi-nested PCR amplification ranging from 10 fg to 10 pg. For the majority of hybridization probes (58/75), the detection limit was 10 fg of template DNA, and for the remaining probes 100 fg, which is an equivalent of less than 1–3 fungal genomes (based on an average fungal genome mass of about 35 fg).

Table 1
Species- genus-specific probes

Probe	Specificity	Acc. no[a], position	Sequence (5′-3′)	Length (nt)	Temp (°C)	GC content (%)	Bead number
A.can1	*A. candidus*	AY373843: 506-526	CAGCCGACCAACCCAACCATT	21	68.38	57.14	4
A.cla1	*A. clavatus*	AY373847: 513-533	CCTGTCGACACCAACCCAATT	21	64.27	52.38	15
A.fla1	*A. flavus*	AY373848: 398-418	GGTCGTCGTCCCCTCTCCGGG	21	73.38	76.19	6
A.fla2	*A. flavus*	AY373848: 504-524	GCGCTTGCCGAACGCAAATCA	21	73.38	57.14	68
A.fla3	*A. flavus*	AY373848: 524-544	AATCTTTTTCCAGGTTGACCT	21	56.81	38.10	75
A.fum1	*A. fumigatus*	AF455542: 529-549	AGCCGACACCCAACTTTATTT	21	59.88	42.86	77
A.fum2	*A. fumigatus*	AF455542: 498-518	TGTCACCTGCTCTGTAGGCCC	21	64.91	61.90	44
A.nid1	*A. nidulans*	AY373888: 455-475	CACCCGCTCGATTAGGGCCGG	21	73.53	71.43	24
A.nid2	*A. nidulans*	AY373888: 479-499	CCAGCCGGCGTCTCCAACCTT	21	71.76	66.67	28
A.nig1	*A. niger*	AY373852: 486-507	ATGCTCTGTAGGATTGGCCGG	21	65.64	57.14	8
A.nig2	*A. niger*	AY373852: 515-535	GACGTTTTCCAACCATTCTT	20	61.24	40.00	11
A.ter1	*A. terreus*	AJ413985: 511-531	GCTTCGTCTTCCGCTCCGTAG	21	65.64	61.90	26

A.ter2	*A. terreus*	AJ413985: 552-572	ACGCATTTATTTGCAACTTGT	21	57.41	33.33	34
Ab.cor1	*Absidia corymbifera*	AY533554: 191-211	GTTGCTGTCATGGCCTTAAAT	21	59.10	42.86	43
Ab.cor2	*Absidia corymbifera*	AY533554: 268-288	GAGCAGCTTGGTTAGTGAGTT	21	56.37	47.62	8
Ab.cor3	*Absidia corymbifera*	AY533554: 336-356	ATGGGACACTACTTGGAGAAA	21	56.24	42.86	11
Ac.str1	*Acremonium strictum*	AY214439: 352-372	TTTCAACCCTCAGGCCCACCC	21	69.57	61.90	41
Ac.str2	*Acremonium strictum*	AY214439: 379-399	GGGAGCGGGCCTGGTTCTGGG	21	74.93	76.19	47
Ac.str3	*Acremonium strictum*	AY214439: 422-442	CGTCCCTCAAATTCAGTGGCG	21	67.14	57.14	30
C.alb1	*C. albicans*	AF217609: 373-394	TGCTTGAAAGACGGTAGTGGT	21	59.79	47.62	30
C.alb2	*C. albicans*	AF217609: 393-413	TAAGGCGGGATCGCTTTGACA	21	67.04	52.38	64
C.alb3	*C. albicans*	AF217609: 437-457	ATTGCTTGCGGCGGTAACGTC	21	68.44	57.14	71
C.dub1	*C. dubiliensis*	AF218993: 130-150	AGGCGGAGATGCTTGACAATG	21	64.75	52.38	41
C.dub2	*C. dubiliensis*	AF218993: 171-191	ATTGCTAAGGCGGTCTCTGGC	21	64.97	57.14	47

(continued)

Table 1 (continued)

Probe	Specificity	Acc. no[a], position	Sequence (5′-3′)	Length (nt)	Temp (°C)	GC content (%)	Bead number
C.dub3	*C. dubiliensis*	AF218993: 190-210	GCGTCGCCCATTTTATTCTTC	21	62.93	47.62	49
C.gla1	*C. glabrata*	AF218994: 201-222	ATCAGTATGTGGGACACGAGC	21	60.01	52.38	38
C.gla2	*C. glabrata*	AF218994: 232-253	CAACTCGGTGTTGATCTAGGG	21	59.60	52.38	13
C.gla3	*C. glabrata*	AF218994: 281-301	TAGGTTTTACCAACTCGGTGT	21	56.33	42.86	67
C.gui1	*C. guillerimondi*	AF218996: 94-114	CTCTTAGTCGGACTAGGCGTT	21	57.79	52.38	89
C.gui2	*C. guillerimondi*	AF218996: 152-172	GCTGTCGACCTCTCAATGTAT	21	56.87	47.62	91
C.gui3	*C. guillerimondi*	AF218996: 192-212	GAATGGTGTGGCGGGATATTT	21	63.01	47.62	92
C.kru1	*C. krusei*	L47113: 329-349	ACGACGTGTAAAGAGCGTCGG	21	64.75	57.14	61
C.kru2	*C. krusei*	L47113: 381-401	GGCCGAGCGAACTAGACTTTT	21	62.07	52.38	83
C.kru3	*C. krusei*	L47113: 421-441	CCGAGAGCGAGTGTTGCGAGA	21	68.47	61.90	62
C.lip1	*C. lipolytica*	AY282524: 96-116	GTACCGCACGGATGGAGGAGC	21	68.35	66.67	54

C.lip2	*C. lipolytica*	AY282524: 128-148	GGGATCGCATTGCTTTCTTGA	21	64.59	47.62	9
C.lip3	*C. lipolytica*	AY282524: 188-208	CCTCCTTCATCCGAGATTACC	21	59.91	52.38	100
C.lus1	*C. lusitaniae*	AY493434: 209-229	CTCCGAAATATCAACCGCGCT	21	65.57	52.38	94
C.lus2	*C. lusitaniae*	AY493434: 230-250	GTCAAACACGTTTACAGCACG	21	59.30	47.62	96
C.lus3	*C. lusitaniae*	AY493434: 250-270	GACATTTCGCCCTCAAATCAA	21	62.22	42.86	98
C.par1	*C. parapsilosis*	AF455530: 321-341	TGAGCGATACGCTGGGTTTGC	21	67.69	57.14	40
C.par2	*C. parapsilosis*	AF455530: 351-371	AGGCGGAGTATAAACTAATGG	21	55.24	42.86	50
C.par3	*C. parapsilosis*	AF455530: 395-415	ACAAACTCCAAAACTTCTTCC	21	55.11	38.10	60
C.tro1	*C. tropicalis*	AF218992: 93-113	ACGCTAGGTTTGTTTGAAAGA	21	56.74	38.10	70
C.tro2	*C. tropicalis*	AF218992: 136-156	AGCGACTTAGGTTTATCCAAA	21	55.81	38.10	81
C.tro3	*C. tropicalis*	AF218992: 156-176	AACGCTTATTTTGCTAGTGGC	21	58.60	42.86	85
Cr.neo1	*Cryptococcus neoformans*	AJ876598: 311-331	AATCTCAATCCCTCGGGTTTT	21	61.01	42.86	17
Cr.neo2	*Cryptococcus neoformans*	AJ876598: 364-384	CGCGACCTGCAAAGGACGTCG	21	72.40	66.67	21

(continued)

Table 1 (continued)

Probe	Specificity	Acc. no[a], position	Sequence (5´-3´)	Length (nt)	Temp (°C)	GC content (%)	Bead number
Cr.nec3	*Cryptococcus neoformans*	AJ876598: 406-426	GGGAAGGTGATTACCTGTCAG	21	58.51	52.38	32
F.oxy1	*Fusarium oxysporum*	AY188919: 356-376	GTGTTGGGACTCGCGTTAATT	21	61.27	47.62	49
F.oxy2	*Fusarium oxysporum*	AY188919: 386-406	CAAATTGATTGGCGGTCACGT	21	65.18	47.62	42
F.sol1	*Fusarium solani*	AJ608989: 331-351	GTCATTACAACCCTCAGGCCC	21	62.88	57.14	94
F.sol2	*Fusarium solani*	AJ608989: 447-467	AGCTAACACCTCGCAACTGGA	21	62.25	52.38	96
F.sol3	*Fusarium solani*	AJ608989: 478-498	GCCATGCCGTAAAACACCCAA	21	67.04	52.38	98
M.muc1	*Mucor mucedo*	AF412289: 388-408	GATGGCCTTTGAGAGTTTACC	21	57.80	47.62	75
M.muc2	*Mucor mucedo*	AF412289: 461-681	ACTGTATGTTCGTAGATGCCC	21	56.71	47.62	77
M.muc3	*Mucor mucedo*	AF412289: 514-534	CGCTTAAAGTCTGCGTGCAAC	21	63.22	52.38	44
M.rac1	*Mucor racemosus*	AF117924: 166-186	GATCTTGAAATCCCTGAAATT	21	54.89	33.33	24
M.rac2	*Mucor racemosus*	AF117924: 196-216	CTGAACTTGTTTAAATGCCTG	21	55.24	38.10	28
M.rac3	*Mucor racemosus, M. plumbeus*	AF117924: 253-273	GACTTTGATGGGGCCTCCCAA	21	67.58	57.14	63

P.chr1	*Penicillium chrysogenum*	AM182189: 488-508	CAACCCGAATTTTTATCCAGG	21	60.52	42.86	81
P.cit1	*Penicillium citrinum*	AM176691: 418-438	CACCCGCTCTAGTAGGCCCGG	21	69.14	71.43	64
P.cit2	*Penicillium citrinum*	AM176691: 456-477	CCAACCTTTAATTATCTCAGGT	22	54.30	36.36	70
P.mar1	*Penicillium marneffei*	AJ853738: 471-491	GGTTGGTCACCACCATATTTA	21	57.20	42.86	92
P.pur1	*Penicillium purpurogenum*	AY373926: 439-459	CGTTGGCCACCCACGATATTT	21	66.04	52.38	71
P.sim1	*Penicillium simplicissimum*	AJ608945: 445-465	CCTCAATCTTTCTCAGGTTGA	21	56.97	42.86	66
R.ory1	*Rhizopus oryzae*	AB126323: 529-549	ATGTGGTAATGGGTCGCATCG	21	65.08	52.38	4
R.ory2	*Rhizopus oryzae*	AB126323: 580-600	GTGTGATTTTCTGTCTGGCTT	21	56.90	42.86	15
R.ory3	*Rhizopus oryzae*	AB126323: 601-621	GCTAGGCAGGAATATTACGCT	21	57.72	47.62	68
Genus-specific probes							
Pan-A/P1[b]	*A. candidus, A. clavatus, A. flavus, A. fumigatus, A. nidulans, A. niger, A. terreus, A. versicolor, A. glaucus, P. chrysogenum, P. citrinum, P. marneffei, P. purpurogenum, P. simplicissimum*	AF455542: 382-403	GCGTCATTGCTGCCCTCAAGC	21	69.59	61.90	43

(continued)

Table 1 (continued)

Probe	Specificity	Acc. no[a], position	Sequence (5¢-3¢)	Length (nt)	Temp (°C)	GC content	Bead number
Pan-A/P2[b]	*A. candidus, A. clavatus, A. flavus, A. fumigatus, A. nidulans, A. niger, A. terreus, A. versicolor, A. glaucus, P. chrysogenum, P. citrinum, P. marneffei, P. purpurogenum, P. simplicissimum*	AF455542: 478-498	TCCTCGAGCGTATGGGGCTT	20	66.35	60.00	45
Pan-Can[b]	*C. albicans, C. dubiliensis, C. glabrata, C. guillerimondi, C. krusei, C. lipolytica, C. lusitaniae, C. parapsilosis, C. tropicalis, C. famata, C. inconspicia, C. kefyr, C. maembranifaciens, C. norvegensis, C. pelliculosa, C. pararugosa, C. sake, C. utilis, C. valida, C. zeylanoides*	AF217609: 232-252	GGGCATGCCTGTTTGAGCGTC	21	69.39	61.90	19
Pan-Fus[b]	*F. oxysporum, F. proliferatum, F. verticilloides*	AY188919: 349-378	GCGTAGTAGTAAAACCCTCG	20	54.39	50.00	54
Pan-Tri1[b]	*T. asahii, T. beigelii, T. inkin, T. cutaneum*	AJ853754: 372-392	GCTCGCCTTAAAAGAGTTAGC	21	57.64	47.62	56
Pan-Tri2[b]	*T. asahii, T. beigelii, T. inkin, T. cutaneum*	AJ853754: 282-302	TTCCGGAGAGCATGCCTGTTT	21	66.35	52.38	2

[a] *Acc no.* accession number

[b] Based on sequence alignments, the spectrum of fungal species recognized by the genus-specific probe is significantly greater than the number of species tested experimentally and presented in the table

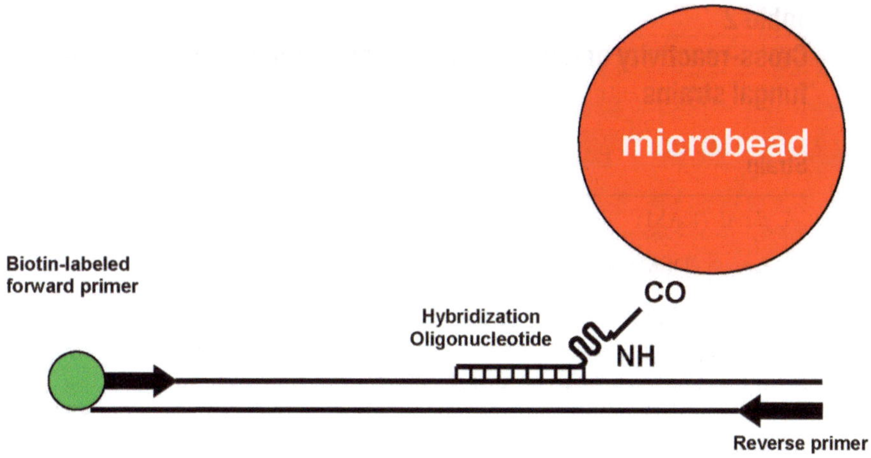

Fig. 1. Principle of the Luminex assay. A double-stranded PCR product is generated by the forward (biotin-labeled) and the reverse primers. The hybridization oligonucleotides (fungal genus- or species-specific probes), which bind to the complementary region on the PCR sense-strand, are coupled via a covalent bond (amide-bond) to the color-coded microbeads. Detection of the PCR amplicons hybridized to the probes is performed via *green fluorescent light* which corresponds to the amount of PCR product (semiquantitative measurement). The *red fluorescent light* identifies the class of microbeads and thus the fungal species present.

The specificity of the Luminex assays was documented by the absence of cross-reactions between hybridization probes and non-homologous DNA. However, the ITS2 rDNA sequences may differ by only a few nucleotides, especially between different species of the genus *Aspergillus*. In these instances, unambiguous identification of different fungal species can be a challenging task. Besides the careful design of hybridization probes, high stringency conditions are used for the hybridization buffer to reduce this problem. Although the occurrence of cross-reactions cannot be entirely excluded (Table 2), they generally occur in one direction only (e.g., *A. terreus* shows some cross-reactivity with one of the probes for *A. flavus* (No 3), but *A. flavus* does not cross-react with *A. terreus* probes]. Hence, unequivocal identification of the fungal species is possible in most instances.

The intra-assay variability of mean fluorescence values (MFI) is in the range of 10% for the majority of probes, but greater fluctuations can be observed in some instances. The Luminex technology therefore only appears to be adequate for semi-quantitative analysis of the fungal targets. In general, the established multiplex assays based on the Luminex technology permit a cost-effective, fast, and reliable identification of clinically relevant fungal pathogens. The implementation of this diagnostic approach can therefore provide the basis for timely initiation of the most appropriate therapy and may thus help preventing unnecessary use of broad-spectrum antifungals.

Table 2
Cross-reactivity of hybridization probes with non-homologous fungal strains

Strain	Hybridization probe
A. flavus (KAM)	A. fum1
A. terreus (DSM826)	A. flav3
Penicillium crysogenum (DSM844)	A. fum2
A. glaucus (KAM)	A. flav3
A. vesicolor (DSM 1953)	A. nid1
C. albicans (DSM 1368)	C. par1
C. parapsilosis (KAM)	C. lus3
Fusarium verticilloides (DSM 62264)	F.oxy2

2. Materials

Reagents used for fungal DNA extraction and semi-nested PCR amplification should be prepared, whenever possible, in a laminar-flow hood using sterile utensils.

2.1. DNA Extraction

1. 15 ml Plastic tubes.
2. PHHV (phocine herpes virus) cell culture supernatant, diluted 1:1,000 using molecular grade water.
3. Red blood cell lysis buffer, sterilized: 10 mM Tris (pH 7.6), 5 mM $MgCl_2$, 10 mM NaCl.
4. Lyticase lysis buffer (LLB); 50 mM Tris (pH 7.6), 1 mM EDTA (pH 8.0), 0.2% 2-mercaptoethanol, 1 U/100 μl recombinant lyticase.
5. Acid-washed glass beads 710–1.180 μm in diameter.
6. MagNA Pure nucleic acid isolation kit I (Roche).
7. Pico Green double-stranded DNA quantification kit (Molecular Probes, Inc., Eugene, OR).
8. MagNA Pure compact instrument (Roche Diagnostics, Basel Switzerland).
9. Fluorescence spectrophotometer.

2.2. Semi-nested PCR Amplification

1. First-round primers: ITS1A forward primer 5′-TTT CAA CAA YGG ATC TCT TGG-3′ Universal ITS4 reverse primer 5′-TCC TCC GCT TAT TGA TAT GCT-3′.

2. Second-round primers, all 5′-end biotinylated: ITS86 I forward primer 5′-TGA ATC ATC GAR TCT TTG AAC G-3′, ITS86 II forward primer 5′-TGA ATC ATC GAG TTC TTG AAC G-3′.
3. Universal ITS4 reverse primer (sequence see above).
4. PCR components for use with AmpliTaq enzyme (ABI); 10× PCR buffer, 25 mM $MgCl_2$, dNTP set and dUTP (100 mM).
5. Heat-labile uracyl-DNA glycosylase (Roche).
6. AmpliTaq DNA polymerase (ABI).
7. Molecular biology grade water.

2.3. Coupling of Capture Probes to x® MAP Luminex Beads

1. x® MAP beads (Luminex, Austin, USA).
2. 0.002% Tween 20 (pH 5.0).
3. Activation buffer $NaH_2PO_4.2H_2O$ pH 6.1.
4. 50 mg/ml *N*-hydoxysulfosuccinimide (NHS; in activation buffer, prepared freshly).
5. 50 mg/ml *N*-(3-dimethylaminodipropyl)–*N′*-ethylcarbodiimide) (EDC; prepared freshly in activation buffer).
6. 50 mM MES (2-(morpholino]ethanesulfonic acid]).
7. 0.5 mg/ml bovine serum albumin solution (in 50 mM MES).
8. 0.1 M MES (pH 5.0).
9. 10 mg/ml EDC (prepared freshly in H_2O).
10. 0.1% Sodium dodecyl sulfate (SDS; in H_2O).
11. TE-buffer: 10 mM Tris–HCl, 1 mM EDTA (pH 8.0).

2.4. Hybridization Assay and Luminex Measurement

1. Hybridization solution (1.5× TMAC): 4.5 M TMAC solution; 0.15% sarcosyl (30% aqueous solution), 75 mM Tris–HCl (pH 8.0), 6 mM EDTA (pH 8.0).
2. Washing solution (1× TMAC): 3 M TMAC solution, 50 mM Tris–HCl (pH 8.0), 4 mM EDTA (pH 8.0).
3. Capture probes: 100 pmol/µl in molecular grade water. (Capture probes sequences are provided in Table 1.)
4. Conjugation buffer: 80 mM Na_2HPO_4, 18 mM KH_2PO_4, 30 mM NaCl.
5. Streptavidin-R-phycoerythrin solution: 1 mg/ml Streptavidin-R-phycoerythrin diluted 1:100 in conjugation buffer.
6. Luminex 100™ apparatus (Luminex Corporation, Austin, Texas). (If another instrument is used for read-out, a number of the indicated components and processes may need to be adapted).
7. 70% Alcohol.
8. 20% Hypochloride solution.

9. Aqua destillata.
10. Luminex® Sheath Fluid (Luminex).
11. Luminex® calibration microspheres (L-100 Cal1 and L-100 Cal2; Luminex).
12. Luminex® control microspheres (L-100 Con1 and L-100 Con2; Luminex).

2.5. Real-Time PCR for the Control of Effective DNA Extraction (Measurement of PHHV Target)

1. 2× Gene Expression Mastermix with UNG (ABI).
2. PHHV forward primer: 5′-GGG CGA ATC ACA GAT TGA ATC-3′.
3. PHHV reverse primer: 5′-GCG GTT CCA AAC GTA CCA A-3′.
4. PHHV probe (5′-FAM-labeled, 3′black hole quencher): 5′-TTT TTA TGT GTC CGC CAC CAT CTG GAT C-3′.
5. Taqman instrument, e.g., ABI 7900 or 7500.

3. Methods

3.1. DNA Extraction

All steps are performed in a laminar flow hood using one-way sterile utensils. Reagents used for DNA extraction are filter-sterilized (see Note 1).

3.1.1. DNA Extraction from Fungal Strains Used as Positive Controls

1. Homogenize a loopful of individual colonies of each fungus culture in 500 μl of LLB and incubat at 37°C for 1 h.
2. After incubation, acid-washed glass beads are added (see Note 2) and the solution is vortexed thoroughly for 2 min (see Note 3).
3. Approximately 400 μl of the supernatant is spiked with 5 μl of PHHV (phocine herpes virus) cell culture supernatant, diluted 1:1,000 using molecular grade water, which serves as a control for efficient DNA extraction.
4. DNA extraction is performed on the MagNA Pure compact instrument using the MagNA Pure compact nucleic acid isolation kit I, as described by the manufacturer (see Note 4).
5. DNA concentrations are determined by the PicoGreen double-stranded DNA quantification kit using the F-2500 fluorescence spectrophotometer.

3.1.2. DNA Extraction from Peripheral Blood Spocimens

1. Transfer 3 ml of peripheral blood, anticoagulated with EDTA, to a 15 ml plastic tube.
2. Fill tube with red blood cell lysis buffer and incubat at room temperature for 15 min.

3. Leukocytes are pelleted by centrifugation for 10 min at 2,000 × *g*.
4. Resuspend pellet in 470 μl LLB.
5. The subsequent steps are identical to the extraction protocol described above.

3.1.3. DNA Extraction from Blood Culture Specimens

1. Transfer 200 μl sample obtained from fungus-positive blood culture to 15 ml tubes and add red blood cell lysis buffer.
2. The subsequent procedure is performed as described above for PB specimens.

3.1.4. DNA Extraction from Plasma Containing White Blood Cells

1. Incubate peripheral blood specimens, anticoagulated with EDTA, at 4°C for at least 4 h to sediment the red blood cells.
2. The entire supernatant (plasma containing white blood cells) is used for DNA extraction.
3. Centrifuged samples at 15,000 × *g* for 10 min.
4. Remove most of the plasma leaving a residual volume of about 100 μl.
5. Add 430 μl of LLB.
6. DNA extraction is performed as described above.

3.1.5. DNA Extraction from Specimens from the Respiratory Tract and Tissue Biopsies

1. Cut solid material into small pieces and homogenize in 430 μl of LLB by vortexing (see Note 3).
2. The ensuing steps are carried out as described above.

To exclude the occurrence of contamination, multiple positive and negative controls have to be included in each run (see Note 5).

3.2. Semi-nested PCR Amplification

For the first round of amplification, the universal ITS1A forward and ITS4 reverse primers are used to amplify a gene sequence covering the complete ITS1, 5.8S, and ITS2 regions as well as portions of the 18S and 28S regions of the rDNA gene. Amplicons containing the entire ITS2 region are generated by a second round of amplification using the 5′ end biotinylated reverse primer ITS4 and two 5′ end biotin-labeled forward primers, ITS86-I and ITS86-II, which hybridize to the 5.8S region of the rDNA gene.

The second forward primer, ITS86-II, is necessary for adequate amplification of *Candida krusei* and *Absidia corymbifera*.

1. Set up PCRs in a total volume of 25 μl containing GeneAmp 10× PCR buffer II, 2.5 mM $MgCl_2$, 5 mM deoxynucleotide triphosphate, dATP, dCTP, dGTP, and a 1:8 ratio of dUTP to dTTP, 400 nM of each primer, 0.25 U heat-labile uracyl-DNA glycosylase (for the first round of amplification only; see Note 6), 2.5 U AmpliTaq DNA polymerase, and molecular biology-grade water (see Note 7).

2. Use 5 μl template rDNA for the first round of amplification.
3. The PCR is performed as follows: 10 min at 37°C (UDG activation); 95°C for 10 min (polymerase activation); 35 cycles of 95°C for 30 s, 55°C for 30 s, and 72°C for 30 s; and a final extension at 72°C for 10 min, followed by cooling at 4°C.
4. Use 3 μl first-round PCR product as template for the second round of amplification (see Note 8).
5. The amplification conditions for the second round of amplification are identical to the first round of PCR, except for omission of the initial UDG activation step.

It is recommended to test the patient samples at least in duplicate reactions. To assess the amplification efficiency of the target sequences by semi-nested PCR and to identify cross-contamination, multiple positive and negative controls have to be included in each run (see Note 9).

3.3. Coupling of Capture Probes to x® MAP Luminex Beads

1. 100 μl or 1.25 million of the appropriate Luminex® beads are used for coupling reactions (see Notes 10 and 11).
2. The bead solutions are supplemented with 0.002% Tween 20, briefly vortexed, and centrifuged for 4 min at 11,000 × *g*.
3. Wash the beads once with 500 μl activation buffer (see Note 12) and resuspend in 192 μl activation buffer by brief vortexing and sonication.
4. Add 24 μl of fresh NHS sodium salt solution and 24 μl of fresh EDC solution (see Note 13).
5. Incubate vials in the dark for 20 min in an ultrasonic bath.
6. Add 0,0002% Tween 20 and centrifuge for 4 min at 11,000 × *g*, carefully remove the supernatant.
7. Wash beads using 500 μl 50 mM MES buffer, 0.002% Tween 20 before resuspending in 250 μl 50 mM MES buffer without Tween.
8. Add 250 μl BSA solutions and incubate the vials for 2 h in the dark under permanent agitation.
9. The samples are supplemented with 0.002% Tween 20 and centrifuged for 4 min at 11,000×*g*, the supernatant is carefully removed. The beads are washed twice with 500 μl 0.1 M MES 0.002% Tween 20, and resuspended in 30 μl 0.1 M MES.
10. Incubate beads in the dark with 10 μl of the capture probes and 10 μl of freshly prepared EDC for 1 h under agitation.
11. After 30 min, add 10 μl of a freshly prepared EDC solution.
12. Wash beads with 1 ml of 0.02% Tween 20 and subsequently with 1 ml of 0.1% SDS.
13. Beads are resuspended in 100 μl TE buffer and stored in the dark at 4°C (stock solution, see Note 14).

3.4. Hybridization Assay and Luminex Measurement

The Luminex assays were optimized for two multiplex reactions (Aspergillus/Emerging fungi [EMF] and Candida/EMF) instead of three reactions described earlier (19). This approach improves the efficiency and throughput in the analysis of patient specimens.

The combination of different oligonucleotide-coupled xMAP® beads used for the Luminex assays is depicted in Tables 3 and 4.

1. The oligonucleotide-coupled xMAP® beads are selected, resuspended by vortexing and sonication, and diluted 1:10 in TE buffer (working concentration).
2. The hybridization mixture contains 0.5 μl of each type of coupled beads (corresponding to approximately 5,000 beads per reaction), 22 μl of hybridization solution (1.5× TMAC; see Note 15), 5 μl of biotinylated PCR amplicons (derived from patients specimens and/or control samples). Add TE buffer to a total volume of 50 μl.
3. Incubate the reaction mixture for 5 min at 95°C in a PCR thermocycler followed by 15 min at 55°C.
4. After hybridization, pellet the beads for 3 min at 1,800×*g* and wash once with 100 μl of washing solution (1× TMAC).
5. Incubate the beads with 100 μl conjugation buffer and 2 μl of streptavidin-R-phycoerythrin solution for 5 min in the dark (see Note 16).
6. The Luminex® 100 apparatus is preset as described by the manufacturer. The detection of PCR-positive/negative signals is performed on the Luminex® 100 apparatus by analyzing 50 beads per capture probe. The raw data are expressed in MFI (mean fluorescence intensity) values. The measured MFI values are subtracted from the background fluorescence signals determined by parallel analysis of control samples (containing all components except PCR amplicons; see Note 17). A signal is regarded as positive if the MFI value is at least twice as high as the background noise.

Multiple positive and negative controls are included in each run to assess the amplification efficiency of the preceding semi-nested PCR amplification of the target sequences and to exclude the occurrence of contamination during the analysis (see Notes 18 and 19).

3.5. Real-Time PCR to Control for Efficient DNA Extraction (PHHV–PCR)

1. Set up PCR in a total volume of 25 μl containing 2× Gene Expression Mastermix with UNG, 900 nM forward primer, 900 nM reverse primer, and 200 nM Taqman® probe.
2. Perform the amplification on a Taqman® instrument (e.g., Taqman® 7900 or Taqman® 7500) using the following cycling conditions: 2 min at 50°C (uracil *N*-glycosylase-mediated degradation of potentially contaminating amplicons containing dUTP), 10 min at 95°C, followed by 50 cycles of 15 s at 95°C and 60 s at 60°C.

Table 3
Aspergillus/emerging fungi assay (30-Plex)

A. can1	A. nig2	P. mar1
A. cla1	A. ter1	F. oxy1
A. fla1	A. ter2	F. oxy2
A. flav2	Pan-A/P1	Pan-Fus
A. flav3	Pan-A/P2	Ac.str1
A. fum1	P. chr1	Ac.str2
A. fum2	P. cit1	Ac.str3
A. nid1	P. cit2	F. sol1
A. nid2	P. sim1	F. sol2
A. nig1	P. pur1	F. sol3

Table 4
Candida/emerging fungi assay (45-Plex)

C. alb1	C. par2	C. tro3	Cr neo3	M. rac2
C. alb2	C. par3	C. lip1	Pan-Tri1	M. rac3
C. alb3	C. kru1	C. lip2	Pan-Tri2	Ab. cor1
C. dub1	C. kru2	C. lip3	R. ory1	Ab. cor2
C. dub2	C. kru3	C. lus1	R. ory2	Ab. cor3
C. dub3	C. gui1	C. lus2	R. ory3	
C. gla1	C. gui2	C. lus3	M. muc1	
C. gla2	C. gui3	Pan-Can	M. muc2	
C. gla3	C. tro1	Cr. neo1	M. muc3	
C. par1	C. tro2	Cr. neo2	M. rac1	

4. Notes

1. All solutions used for DNA extraction are filtered through 0.2 μm sterile filters (Corning; Corning, Incorporated, Germany) to remove potentially present fungal spores.
2. For the addition of acid-washed glass beads, it is recommended to use half of the volume of a 1 ml sterile pipette tip.

3. To facilitate strong agitation and to intensify mechanical disruption of the fungal cell walls, it is recommended to use the Disruptor Genie machine (Scientific Industries, New York, USA) or a similar device.

4. For optimal time management, it is suggested to set up the MagNa Pure® instrument according to the manufacturer's instructions during the incubation time with LLB (1 h).

5. To determine the potential occurrence of contamination during fungal DNA extraction, at least one control sample containing all required reagents but no fungal or patient material is prepared (extraction control). To ensure adequate performance of the DNA isolation, each patient sample is spiked prior to DNA extraction with 5 μl PHHV (phocine herpesvirus) cell culture supernatant, diluted 1:1,000 with molecular grade water. Successful DNA extraction is determined by real-time PCR amplification of a PHHV target sequence and the presence of quantitative results within the expected range.

6. The use of heat-labile uracil-DNA glycosylase (UNG) prevents contamination with PCR products by digesting PCR amplicons containing incorporated dUTP. A UNG-digestion step is therefore included in the first round of the semi-nested PCR. Since the amplicons containing dUTP obtained from the first PCR round are used as template for the second round of amplification, the UNG step has to be omitted in this step to prevent digestion of the PCR template.

7. Different commercial products of sterile water were carefully evaluated for their eligibility for molecular fungal diagnostics. A number of preparations were found to contain traces of fungal DNA which resulted in false-positive PCR findings. The molecular-grade water provided by 5 PRIME was found to meet the requirements of sterility for molecular fungus analysis.

8. Handling PCR products required in (semi)-nested PCR analysis is a potential risk for contamination. Working in a designated area or in a PCR-hood, and using dedicated pipettes, sterile filter-tips and other consumables helps preventing cross-contamination.

9. DNA specimens from at least three different fungal strains are used as positive controls for semi-nested PCR and for the Luminex assays. Duplicate samples of molecular biology grade water are used as negative controls and for the establishment of blank values permitting subtraction of background noise from the measured MFI readings.

10. To protect the color-coded Luminex® xMap beads from light, it is recommended to use Eppendorf 1.5 ml tubes "amber" (Eppendorf, Hamburg, Germany) for storage and handling of the beads.

11. To facilitate economic use of Luminex® xMAP beads, it is recommended to implement the same types of uncoupled beads in both assays. A scheme for optimal distribution is provided in Table 1.
12. Removal of the supernatant during the coupling procedure of Luminex® xMAP beads is a crucial step and has to be carried out carefully to avoid aspiration and thus loss of beads.
13. To ensure high activity, NHS and EDC solutions have to be prepared immediately before use.
14. Coupled Luminex® xMap beads are stable for ~1 year at 4°C. Low MFI values can be a sign of degraded capture probes.
15. The use of high concentrations of TMAC, an ammonium salt agent that increases the stringency of hybridization, is essential for the discrimination of fungal target sequences differing by a few nucleotides only. This compound renders the efficiency of oligonucleotide probe hybridization dependent on the length of the probe rather than on the base composition.
16. Incubation with streptavidine/phycoerythrine can be performed in the Luminex® device.
17. For analysis of the Luminex (raw) data and calculation of the final results, the MFI values have to be formated as an Excel file.
18. If the Luminex readings indicate negative results for positive controls, it is recommended to first visualize the PCR products by gel-electrophoresis to exclude an error during PCR amplification.
19. In case negative controls provide positive results, indicating contamination, the tests have to be repeated. To determine whether contamination has occurred during the semi-nested PCR, it is recommended to visualize the PCR products by gel electrophoresis. In case of contamination during the PCR, the negative control (molecular-grade water) shows a positive signal upon electrophoresis.

References

1. Walsh TJ et al (2008) Treatment of aspergillosis: clinical practice guidelines of the Infectious Diseases Society of America. Clin Infect Dis 46:327–360
2. Richardson M, Lass-Florl C (2008) Changing epidemiology of systemic fungal infections. Clin Microbiol Infect 14(Suppl 4):5–24
3. Cornely OA (2008) Aspergillus to Zygomycetes: causes, risk factors, prevention, and treatment of invasive fungal infections. Infection 36:296–313
4. Caston-Osorio JJ et al (2008) Epidemiology of invasive fungal infection. Int J Antimicrob Agents 32(Suppl 2):S103–S109
5. Lass-Florl C (2009) The changing face of epidemiology of invasive fungal disease in Europe. Mycoses 52:197–205
6. Malani AN, Kauffman CA (2007) Changing epidemiology of rare mould infections: implications for therapy. Drugs 67:1803–1812
7. Marr KA et al (2002) Aspergillosis, pathogenesis, clinical manifestations, and therapy. Infect Dis Clin North Am 16:875–894, vi
8. Morace G, Borghi E (2010) Fungal infections in ICU patients: epidemiology and the role of diagnostics. Minerva Anestesiol 76: 950–956

9. Lau A et al (2007) Development and clinical application of a panfungal PCR assay to detect and identify fungal DNA in tissue specimens. J Clin Microbiol 45:380–385
10. Leaw SN et al (2006) Identification of medically important yeast species by sequence analysis of the internal transcribed spacer regions. J Clin Microbiol 44:693–699
11. Landlinger C et al (2008) Identification of fungal species by fragment length analysis of the internally transcribed spacer 2 region. Eur J Clin Microbiol Infect Dis 28(6):613–622
12. Gharizadeh B et al (2004) Identification of medically important fungi by the Pyrosequencing technology. Mycoses 47:29–33
13. Montero CI et al (2008) Evaluation of Pyrosequencing technology for the identification of clinically relevant non-dematiaceous yeasts and related species. Eur J Clin Microbiol Infect Dis 27:821–830
14. Leinberger DM et al (2005) Development of a DNA microarray for detection and identification of fungal pathogens involved in invasive mycoses. J Clin Microbiol 43:4943–4953
15. Spiess B et al (2007) DNA microarray-based detection and identification of fungal pathogens in clinical samples from neutropenic patients. J Clin Microbiol 45:3743–3753
16. Das S et al (2006) DNA probes for the rapid identification of medically important Candida species using a multianalyte profiling system. FEMS Immunol Med Microbiol 46: 244–250
17. Diaz MR, Fell JW (2004) High-throughput detection of pathogenic yeasts of the genus trichosporon. J Clin Microbiol 42:3696–3706
18. Zeng X et al (2007) Reverse line blot hybridization assay for identification of medically important fungi from culture and clinical specimens. J Clin Microbiol 45:2872–2880
19. Landlinger C et al (2009) Species-specific identification of a wide range of clinically relevant fungal pathogens by the Luminex xMAP™ technology. J Clin Microbiol 47(4):1063–1073
20. Landlinger C et al (2010) Diagnosis of invasive fungal infections by a real-time panfungal PCR assay in immunocompromised pediatric patients. Leukemia 24:2032–2038

Chapter 10

Applied Gene Histopathology: Identification of *Fusarium* Species in FFPE Tissue Sections by In Situ Hybridization

Yoichiro Okubo, Minoru Shinozaki, Megumi Wakayama, Haruo Nakayama, Daisuke Sasai, Takao Ishiwatari, Tetsuo Nemoto, Tochigi Naobumi, and Kazutoshi Shibuya

Abstract

Although accurate and rapid diagnosis of fusariosis is now required, morphological similarities among molds make it difficult to histologically differentiate *Fusarium* spp. from other molds. In this chapter, we present our in situ hybridization (ISH) technique as a valuable tool to identify *Fusarium* spp. and emphasize the usefulness of the technique.

Key words: In situ hybridization, *Fusarium* species, Invasive fungal infection, Immunocompetent individual, Peptide nucleic acid, Formalin-fixed and paraffin-embedded tissue section, 28S rRNA

1. Introduction

Invasive fungal infection (IFI) is now more common and has a greater impact in developed countries (1, 2). In particular, *Fusarium* species are widely distributed and are regarded as an emerging and life-threatening fungal pathogen in immunocompromised individuals (Fig. 1) (3). Additionally, it has been shown that *Fusarium* spp. are resistant to most antifungal agents (4). Therefore, the accurate and rapid diagnosis of the infection, especially identification in regard to *Aspergillus* spp., is now needed to improve the outcome of treatment for seriously debilitating conditions because the fungus has little 1, 3-Beta-D-glucan, which is regarded as a target molecule of the antifungal candin class. However, morphological similarities among molds make it difficult to histologically differentiate

Louise O'Connor and Barry Glynn (eds.), *Fungal Diagnostics: Methods and Protocols*, Methods in Molecular Biology, vol. 968, DOI 10.1007/978-1-62703-257-5_10, © Springer Science+Business Media New York 2013

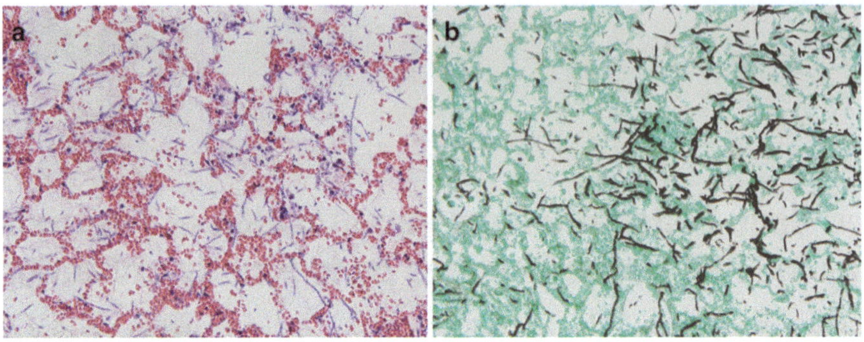

Fig. 1. Pulmonary lesion of mouse 3 days after intratracheal infection of *Fusarium solani* (TIMM1303). An extensive mold growth is observed in alveoli. Molds show hyphal growth with uniform width and partial dichotomous branching **(a)** Hematoxylin and Eosin double stain, 400×, **(b)** Grocott's stain, 400×).

Fusarium spp. from other molds (5). To solve this problem, we previously reported a method of in situ hybridization (ISH) using peptide nucleic acid (PNA) as the probe targeting the 28S rRNA of *Fusarium* spp. to identify the fungus in formalin-fixed and paraffin-embedded (FFPE) tissue sections (6). Since FFPE sections are widely employed in routine preparations for surgical and anatomical pathology (7), our proposed ISH technique may be utilized in many medical institutes. Furthermore, patients can avoid additional diagnostic procedures by analysis of the stored paraffin-embedded block. Although further studies are needed to establish ISH with PNA probes as an accurate and rapid diagnostic procedure for fusariosis, we wish to emphasize the usefulness of our technique involving ISH with PNA probes (Fig. 2).

2. Materials

In our proposed ISH technique, most of the materials have been regarded as the standard tissue preparation method in medical institutes. In the present section, we describe exceptional materials which require several adjustments.

1. Glass slide: Silane-coated glass slide (Dako Japan, Tokyo, Japan) is used.
2. 1 mM Concentration of ethylenediaminetetraacetic acid (EDTA) (Dojindo Laboratories, Kumamoto, Japan) buffer: Add 0.37 g of EDTA (SIGMA product E-5134) to 1 l of distilled water. Adjust pH to 8.0 using 1.0 M NaOH.
3. 0.1 M Concentration of phosphate-buffered saline (PBS): Dissolve 29.01 g of $Na_2HPO4/12H_2O$ and 2.96 g of $NaH_2PO_4/2H_2O$ in 100 ml of distilled water (0.1 M concentration of PBS). And then, dissolve 100 ml of 0.1 M concen-

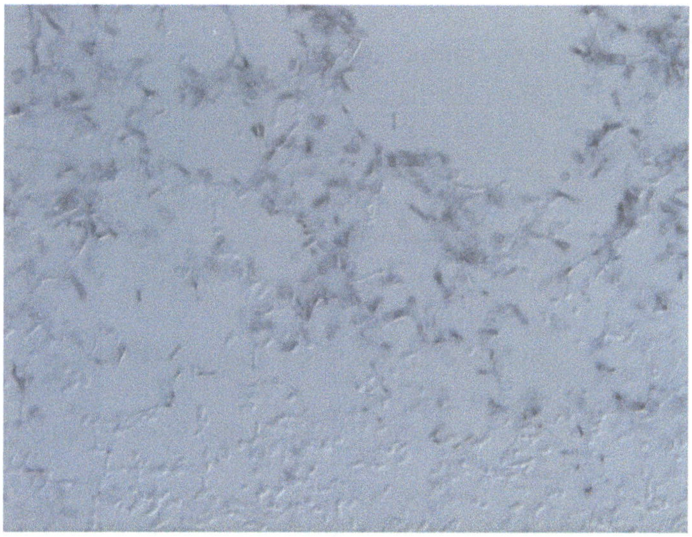

Fig. 2. In situ hybridization with peptide nucleic acid probe in infected mice. The tissue sections were hybridized with *Fusarium* spp. peptide nucleic acid probe. Strong positive signals against 28S rRNA of *Fusarium* spp. were observed in lung tissue from mice infected with *Fusarium solani* (TIMM1303) (400×).

tration of PBS in 900 ml of distilled water (0.01 M concentration of PBS). Finally, add 8.5 g of NaCl to the PBS.

4. 10 μg/ml Concentration of proteinase K: Dissolve 20 mg/ml concentration of proteinase K (Nippon Gene Co., Ltd., Tokyo, Japan) in 30 ml of 0.01 M concentration of PBS.
5. 4% Paraformaldehyde: Add 8 g paraformaldehyde powder in 180 ml of distilled water and heat to 60° to dissolve. And then, add 40 μ/l of 10 N NaOH. After paraformaldehyde is completely dissolved, cooled the solution with ice water and add 20 ml of 10× PBS. Check that pH = 7.4 and filter 4% paraformaldehyde through a 0.45 μm filter.
6. 50% Formamide/2× standard saline citrate (SSC): Make a 1:10 dilution from 20× SSC stock (3 mol/l NaCl, 0.3 mol/l sodium citrate)with sterile deionized water and add 1:1 formamide.
7. 1.0% Periodic acid: Dissolve 0.5 g of periodic acid (Wako Pure Chemical, Osaka, Japan) in 50 ml of water.

3. Methods

3.1. Preparation of FFPE Tissue Block (see Note 1)

1. Fix freshly dissected tissues in 10% formalin for 24–48 h at room temperature.
2. Cut the tissues to proper size, which can be 20 × 20 × 5 mm, and then put the tissues into embedding cassettes.

3. If necessary fix the tissues in 10% formalin, again at room temperature.
4. Rinse the tissues with running tap water for 10 min at room temperature.
5. Dehydrate the tissues through 70%, 80%, and 95% ethanol, 45 min each, followed by 3 times with 100% ethanol for 1 h each at room temperature.
6. Immerse the tissues in 100% xylene 3 times for 1 h each at room temperature.
7. Immerse the tissues in paraffin 4 times for 1 h each at room temperature.
8. Embed the tissue in a paraffin block. The paraffin tissue block can be stored at room temperature.

3.2. In Situ Hybridization

1. Prepare a 3-μm thick paraffin section and mount it on silane-coated glass slides (see Note 2).
2. Dry the slides for 60 min at 60°C (see Note 3).
3. Immerse the slides in 100% xylene 3 times for 10 min each at room temperature.
4. Immerse the slides in a 100% ethanol bath 3 times for 5 min each at room temperature.
5. Immerse the slide in sterile water 3 times at room temperature (see Note 4).
6. Immerse the slide in a 1 mM concentration of EDTA buffer (pH 8.0) for 20 min at 98°C.
7. Cool the slides for 20 min at room temperature.
8. Immerse the slides in PBS 3 times for 5 min each at room temperature (see Note 5).
9. Immerse the slides in a 10 μg/ml concentration of proteinase K (Nippon Gene Co., Ltd., Tokyo, Japan) for 10 min at 37°C (see Note 6).
10. Immerse the slides in 4% paraformaldehyde for 10 min at room temperature.
11. Immerse the slides in PBS 3 times for 5 min each at room temperature.
12. Immerse the slides in sterile water 2 times for 15 s at room temperature.
13. Dehydrate the tissue section on the slides in graded ethanol.
14. Air-dry the slides for 10 min.
15. Apply 20 μl of 1 μg/ml concentration of PNA probe dissolved in hybridization buffer (NIPPON GENE CO., LTD. Toyama, Japan) and mount the coverslip on the slides (see Note 7).

16. Co-denature the tissue section and PNA probe by placing the slides on a heat block (94°C) for 10 min.
17. Cool the slides for 5 min on a refrigerant.
18. Pour 50% formamide/2× SSC into the bottom of a humidity chamber and place the chamber in the incubator (see Note 8).
19. Hybridize the PNA probe for 90 min at 56°C.
20. Remove coverslips by immersing the slides in a 2× SSC solution and allowing the cover slips to fall from the slides.
21. Immerse the slides in 0.2× SSC 2 times for 20 min each at 56°C.
22. Immerse the slides in 1.0% periodic acid for 10 min at room temperature (see Note 9).
23. Wash the slides in PBS 3 times for 5 min each at room temperature.
24. Incubate with anti-FITC antibody (Roche Diagnostics K.K., Tokyo, Japan) for 60 min.
25. Immerse the slides in PBS 3 times for 5 min each at room temperature.
26. Incubate with a horseradish peroxidase-labeled polymer solution (Nichirei Biosciences, Inc., Tokyo, Japan) for 30 min.
27. Immerse the slides in PBS 3 times for 5 min each at room temperature.
28. Incubate with 3,3′-Diaminobenzidine, tetrahydrochloride (DAB) (Dojindo Laboratories, Kumamoto, Japan) in the presence of H_2O_2, nickel, and cobalt ions for 4 min.
29. Quickly immerse the slides in a 100% ethanol bath 3 times at room temperature.
30. Quickly immerse the slides in 100% xylene 3 times at room temperature.
31. Mount coverslips on the slides with mounting medium (e.g., Malinol, Muto Pure Chemicals, Tokyo, Japan).

4. Notes

1. In our proposed ISH technique, the FFPE tissue block is used as the material for analysis. This has been regarded as the standard tissue preparation method in medical institutes.
2. The silane-coated glass slides are essential to our proposed ISH. Its superior adhesion can prevent the tissue sections from falling off the slides.

3. Optimal dehydration is essential to remove water from beneath the tissue sections and it makes the tissue sections adhere better to the slides. In our laboratory we use a paraffin oven (PM-401; SAKURA SEIKI Co. Ltd. Tokyo, Japan).
4. This procedure is essential to prevent ribonuclease (RNase) contamination. RNases are very stable and easily destroy RNA. Great care should be taken to create and maintain an RNase-free environment. The sterile distilled water used to prepare solutions and buffers was treated with 0.1% (v/v) diethyl pyrocarbonate (DEPC; Sigma). All tips, tubes, and DEPC-treated water were autoclaved at 121.1°C for 20 min. Furthermore, to prevent RNase contamination from the surface of the skin, disposable gloves should be worn at all times.
5. PBS is a buffer solution used in a variety of biological applications. Since the osmolarity and ion concentrations of PBS match those of the human body, it is isotonic and nontoxic to cells. Furthermore, PBS helps to maintain a constant pH. In our laboratory, PBS is adjusted to pH 7.6 and a concentration of 0.01 M.
6. Proteinase K digestion can increase the sensitivity of ISH due to the partial removal of formalin crosslinking. However, it is well known that insufficient digestion results in a poor hybridization signal, and over digestion more or less destroys tissue morphology. Since the optimal concentration of proteinase K varies depending upon the tissue type, size, and length of formalin-fixation, the concentration of proteinase K should be adjusted in each case.
7. The design of the PNA probes for *Fusarium* spp. (N-terminus—GAT GAT CAA CCA AGC CCA) was derived from a comparison of 28S rRNA genes in the GenBank database. The N-terminus of the oligomer was conjugated to fluorescein isothiocyanate via a double aminoethoxyethoxyacetate linker (Fasmac Co., Ltd., Kanagawa, Japan).
8. 50% Formamide/2× SSC is essential to maintain humidity and prevent the slides from drying. A layer of chromatography paper or any absorbent paper should be placed on the bottom of the humidity chamber.
9. Endogenous peroxidase activity is found in many tissues and this may cause a high background signal. Therefore, inactivation of endogenous peroxidase is essential to increase the specificity and sensitivity of the procedure. In formalin-fixed tissues, this activity can be quenched by 1.0% periodic acid. A solution of H_2O_2 can also be used.

Acknowledgments

This work was supported by the Health Science Research Grants for Research on Emerging and Re-emerging Infectious Diseases (H22-Shinkou-Ippan-8 and H23-Shinkou-Ippan-018), Measures for Intractable Diseases (H20-Nannchi-Ippann-35) from the Ministry of Health, Labor and Welfare of Japan, a grant from the Strategic Basis on Research Grounds for Non-governmental Schools at Heisei 20th from the Ministry of Education, Culture, Sports, Science and Technology of Japan to KS, a grant from the Strategic Research Foundation Grant-aided Project for Private schools at Heisei 23rd from Ministry of Education, Culture, Sports, Science and Technology of Japan to KS, 2011-2015, a Toho Universi ty project grant #23-19, 21, and 28 and #24-11, 16, 27, and 28, Yokohama Foundation for Advancement of Medical Science grant to YO and MS, Dr Yanase's grant of Toho University Medical School to YO, and Grant-in-Aid for Young Scientists B from the Ministry of Education, Culture, Sports, Science and Technology of Japan (KAKENHI, #24790364) to YO.

References

1. Nakayama H, Shibuya K, Kimura M et al (2010) Histopathological study of candidal infection in the central nervous system. Nihon Ishinkin Gakkai Zasshi 51:31–45
2. Shibuya K, Coulson WF, Naoe S (2002) Histopathology of deep-seated fungal infections and detailed examination of granulomatous response against cryptococci in patients with acquired immunodeficiency syndrome. Nihon Ishinkin Gakkai Zasshi 43:143–151
3. Nelson PE, Dignani MC, Anaissie EJ (1994) Taxonomy, biology, and clinical aspects of Fusarium species. Clin Microbiol Rev 7:479–504
4. Guinea J, Peláez T, Recio S et al (2008) In vitro antifungal activities of isavuconazole (BAL4815), voriconazole, and fluconazole against 1,007 isolates of zygomycete, Candida, Aspergillus, Fusarium, and Scedosporium species. Antimicrob Agents Chemother 52:1396–1400
5. Shinozaki M, Okubo Y, Nakayama H et al (2009) Application of in situ hybridization to tissue sections for identification of molds causing invasive fungal infection. Nihon Ishinkin Gakkai Zasshi 50:75–83
6. Shinozaki M, Okubo Y, Sasai D et al (2011) Identification of Fusarium species in formalin-fixed and paraffin-embedded sections by in situ hybridization using peptide nucleic acid probes. J Clin Microbiol 49:808–813
7. Okubo Y, Shinozaki M, Yoshizawa S et al (2010) Diagnosis of systemic toxoplasmosis with HIV infection using DNA extracted from paraffin-embedded tissue for polymerase chain reaction: a case report. J Med Case Reports 4:265

Chapter 11

Application of Chip-Based Flow Cytometry for Amphotericin B and Fluconazole Susceptibility Testing on *Candida* Strains

Orsolya Bouquet, Béla Kocsis, Ferenc Kilár, and Ildikó Kustos

Abstract

Chip-based flow cytometry is a rather new method that offers an easy, fast opportunity for examination of yeasts, such as *Candida* cells. In our study cell-chip technology was tested with ATCC *Candida* strains to determine their viability and susceptibility against antifungal agents, amphotericin B and fluconazole. We found this technology to be suitable for the detection of *Candida* cells, for the differentiation between dead and living cells, and for the determination of amphotericin B and fluconazole susceptibility of different *Candida* strains (Bouquet et al., Mycoses 55:e90–e96, 2012).

Key words: Candida, Lab-on-a-chip, MIC, Amphotericin B, Fluconazole, Flow cytometry, Agilent 2100 bioanalyzer

1. Introduction

Flow cytometry (FC) is a well-known, versatile method used for the examination of fungal cells, even for susceptibility testing (2–4). Different methods are available to determine minimum inhibitory concentration (MIC) from the FC results in the literature (5, 6).

Chip-based flow cytometry is a new application that is not as complex as the conventional flow cytometry but also offers some advantages (7, 8). These advantages include increased sensitivity, easy management, and low sample consumption. On the other side only fluorescently labeled cells can be analyzed and the experimental conditions are set, less variable. The number of microbiological applications of lab-on-a chip systems, including chip-based flow cytometry, are growing (1), (9–11).

Louise O'Connor and Barry Glynn (eds.), *Fungal Diagnostics: Methods and Protocols*, Methods in Molecular Biology, vol. 968, DOI 10.1007/978-1-62703-257-5_11, © Springer Science+Business Media New York 2013

We used this method to determine susceptibility of *Candida albicans* and *non-albicans* strains to amphotericin B and fluconazole. The measurements were carried out according to the manufacturer's instruction. The most important parts of this application are the sample preparation and evaluation of data. With the chip methodology determination of MIC values were based on the differentiation between living and dead cells. *Candida* cells were incubated in the presence of different amounts of antifungal agents. Depending on the concentration of the antifungals the multiplication of yeast cells were inhibited by different extent. After incubation all samples underwent heat treatment as Sytox Green, the fluorescent dye utilized in our system labels only dead cells. The higher the concentrations of antifungal agents were the lower fluorescence signals could be detected. The MIC value was described as the concentration where the ratio of the living cells substantially (min. 50%) decreased compared with the growth control (12).

2. Materials

1. Sabouraud dextrose agar plates were used for maintenance of fungal cells.
2. Medium for multiplication—RPMI 1640 broth (Sigma-Aldrich Chemie GmbH, Steinheim, Germany).
3. Sytox Green (Molecular Probes, Eugene, Oregon, USA) non-permeable nucleic acid binding stain was used as fluorescent label (5 mM solution in DMSO). The approximate fluorescence excitation/emission maxima of Sytox Green are 504/523 nm (see Note 1).
4. Antifungal agents: fluconazole and amphotericin B. Amphotericin B—in the form of Fungizone® powder for injection (Bristol-Myers Squibb, Epemon, France) Stock solution (1 mg/ml in distilled water) was prepared before the experiments. Fluconazole was used in the form of Mycosyst 2 mg/ml infusion (Richter-Gedeon, Budapest, Hungary).
5. Agilent 2100 Bioanalyzer with the Cell chip kit that contains reagents (Priming Solution, Focusing Dye, Cell Buffer) and chips. The Agilent 2100 Bioanalyzer is equipped with a red laser and a blue LED. The red laser has an excitation maximum at approx. 630 nm (detection maximum at 680 nm), whereas the excitation maximum of the blue LED is at 470 nm (detection maximum at 525 nm).
6. Benchtop centrifuge.

3. Methods

3.1. Preparation of Samples

1. Inoculate fungal strains into 1 ml RPMI 1640 medium and incubate for 2 h at 30°C, until the optical density of the suspensions reaches $A = 0.4$ at 600 nm (see Note 2).
2. Prepare macrodilution series of each antifungal agent in sterile tubes using 1 mL of RPMI 1640 medium as diluent (see Note 3) as outlined in the following steps.
3. The concentration of the antifungal agent in the first tube is set: 10 μg/ml of amphotericin B or 62 μg/ml of the fluconazole. To achieve the concentration of amphotericin B in the first tube, add 20 μl of the 1 mg/ml amphotericin B stock solution and bring the volume to 2 ml with the RPMI medium. To achieve a 64 μg/ml concentration of fluconazole in the first tube, add 32 μl of the 2 mg/ml fluconazole infusion solution and bring the volume to 2 mL with the RPMI media.
4. Vortex this first tube for a few seconds and then prepare the set of serial dilutions (1:1) transferring 1 mL at each stage.
5. Add 10 μl of incubated fungal suspensions to each tube.
6. Prepare positive control: Add 10 μl of incubated fungal suspension to a tube containing 1 ml of RPMI broth without antifungal agent.
7. Prepare negative control: This tube should contain only 1 ml of RPMI broth without antifungal agent or fungi.
8. Incubate all samples for 24 h at 30°C.

3.2. Label Cells

1. Transfer 350 μl of each samples (including controls) to Eppendorf tubes. From the positive control make a duplicate.
2. Place the samples, except one aliquot of the positive control, in an Eppendorf heater or water-bath at 100°C for 10 min. Allow samples cool down at room temperature.
3. Add 0.5 μl of Sytox Green to each tube. Vortex gently the samples and incubate them at room temperature for 30 min (see Note 4).

3.3. Chip Measurements

1. Let the cell-chip reagents reach room temperature.
2. Centrifuge 100 μl of heat-treated and stained samples at 3,000×g for 10 min (see Note 5). Resuspend pellets in the same amount (100 μl) of Cell Buffer. Vortex sample vigorously until the pellets solved properly.
3. Prepare the microchip: prime with 10 μl of the Priming Solution, filling the channels of the chip. Then pipette 10 μl of the Focusing Dye and 30 μl of the Cell Buffer to the proper wells on the chip as outlined in the manufacturers instructions.
4. Pipette 10 μl of each sample into a sample well and start measurements. Use antibody staining program of the Agilent 2100 Expert program (see Note 6).

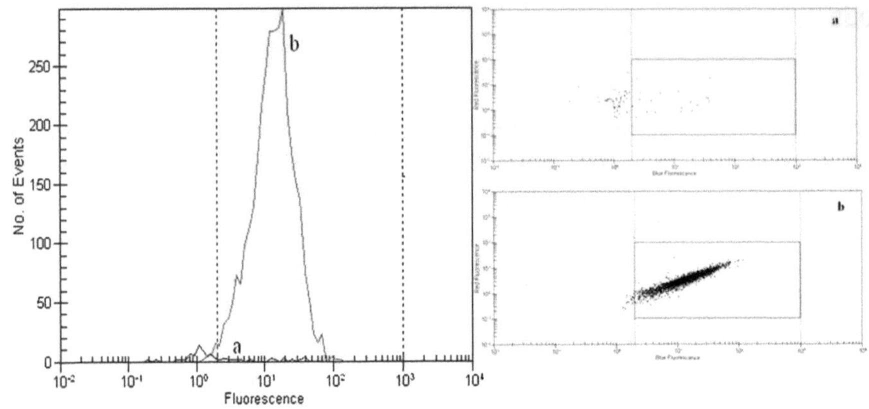

Fig. 1. The cell-chip results of control samples of *C. albicans* are presented. Histogram and dot plot *a* shows the dead cells (non-heat-treated positive control), while figure *b* demonstrated the total amount of cells in the sample (heat-treated positive control).

3.4. Evaluation

1. Set the proper gates of *Candida* cells using the results of the negative and positive controls (see Note 7).
2. The heat treated aliquot of the positive control determines the total amount of cells (Fig. 1b), while the non-treated aliquot of the positive control shows the amount of dead cells in the sample (Fig. 1a). The difference of these equals the amount of living cells. MIC is the concentration of the antifungal agent where the ratio of the living cells substantially (min. 50%) decreased comparing with the growth control (Figs 2 and 3).

4. Notes

1. Several other fluorescent dyes can be used that are able to differentiate between living and dead cells and compatible with the instrument.
2. If the stock suspensions of fungi are too concentrated it affects outcome resulting in falsely high MIC.
3. Make parallel series and repeat measurements 3 times to get reproducible results.
4. Sytox Green is a fluorescent dye, thus protect stained samples from light.
5. Samples might be washed after staining step. In case of Sytox Green it is not obligatory but in case of other dyes it may be.

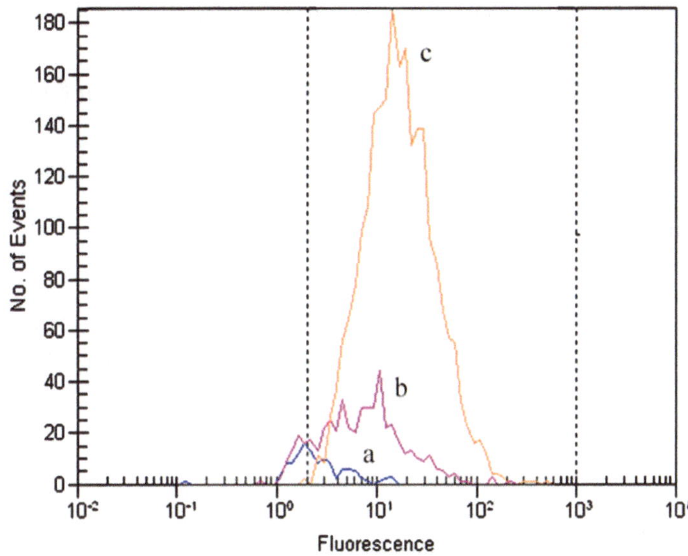

Fig. 2. Results of an amphotericin B macrodilution series on *C. albicans ATCC 90028* are presented. The histograms show fluorescence intensity detected at the 510–540 nm region versus the number of events. Patterns obtained at different concentrations of the amphotericin B are presented: 0.312 μg/ml in histogram *a*, 0.156 μg/ml in histogram *b*, and 0.078 μg/ml in histogram *c*. Concentrations higher than the MIC value (histogram *a* and *b*) inhibit the growth of fungal cells, thus very low fluorescent signals could be detected. At concentrations lower than the MIC (histogram *c*) fungal cells were able to multiplicate thus the histogram shows higher fluorescent values.

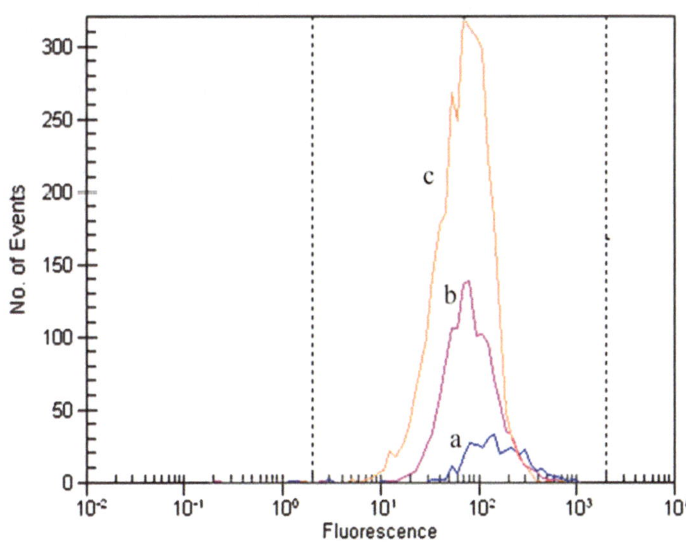

Fig. 3. Cell chip results of a fluconazole macrodilution series on *C. albicans* are demonstrated. Patterns obtained at different concentrations of the fluconazole are presented: 1 μg/ml in histogram *a*, 0.5 μg/ml in histogram *b*, and 0.25 μg/ml in histogram *c*. Increasing concentrations of fluconazole had increasing inhibitory effect on multiplication of *Candida* cells. In the case of fluconazole even at concentrations higher than MIC value (histogram *a* and *b*) much higher fluorescent signals could be detected. This might be explained by the fact that amphotericin B has fungicidal, while fluconazole only fungistatic effect on *Candida* cells.

6. General assay can be used too, especially if other dyes are used.
7. Gates should be set to exclude debris but contain the signals of fungal cells.

Acknowledgments

Funded by: GVOP; Grant Number: 3.2.1-0168; RET; Grant Number: 008–2005 OTKA; Grant Number: NI-68863.

References

1. Bouquet O et al (2012) Amphotericin B and fluconazole susceptibility of Candia species determined by cell-chip technology. Mycoses 55:e90–e96.doi:10.1111/j.1439-0507.2011.02118.x
2. Kirk SM et al (1997) Rapid susceptibility testing of Candida albicans by flow cytometry. J Clin Microbiol 35:358–363
3. Pina-Vaz C et al (2005) Comparison of two probes for testing susceptibilities of pathogenic yeasts to voriconazole, itraconazole, and caspofungin by flow cytometry. J Clin Microbiol 43:4674–4679
4. Wenisch C et al (2001) Antifungal susceptibility testing of fluconazole by flow cytometry correlates with clinical outcome. J Clin Microbiol 39:2458–2462
5. Ramani R, Ramani A, Wong SJ (1997) Rapid flow cytometric susceptibility testing of Candida albicans. J Clin Microbiol 35:2320–2324
6. Mitchell M, Hudspeth M, Wright A (2005) Flow cytometry susceptibility testing for the antifungal caspofungin. J Clin Microbiol 43:2586–2589
7. Andersson H, van den Berg A (2003) Microfluidic devices for cellomics: a review. Sens Actuators B Chem 92:315–325
8. Preckel T et al (2002) Detection of cellular parameters using a microfluidic chip-based system. J Assoc Lab Autom 4:85–89
9. Sakamoto C, Yamaguchi N, Nasu M (2005) Rapid and simple quantification of bacterial cells by using a microfluidic device. Appl Environ Microbiol 71:1117–1121
10. Palkova Z et al (2004) Single-cell analysis of yeast, mammalian cells, and fungal spores with a microfluidic pressure-driven chip-based system. Cytometry A 59A:246–253
11. Gerdts G, Luedke G (2006) Fish and chips: marine bacterial communities analyzed by flow cytometry based on microfluidics. J Microbiol Methods 64:232–240
12. Ramani R, Chaturvedi V (2000) Flow cytometry antifungal susceptibility testing of pathogenic yeasts other than Candida albicans and comparison with the NCCLS broth microdilution test. Antimicrob Agents Chemother 44:2752–2758

Chapter 12

Surface Plasmon Resonance Genosensor for the Detection of *Fusarium culmorum*

Michelangelo Pascale, Francesco Zezza, and Giancarlo Perrone

Abstract

Surface plasmon resonance (SPR)-based DNA biosensors have been shown to be rapid, label-free, and selective tools for the detection of PCR products. Here, we describe an SPR sensor based on DNA hybridization for the detection of *Fusarium culmorum*, a fungal pathogen of wheat. A 0.57 kb DNA fragment of *F. culmorum* was amplified by specific primers, and a 25-mer oligonucleotide probe was selected within the sequence of the PCR amplicon. The biotin-labeled probe was immobilized on a streptavidin sensor chip and tested for biospecific interaction with PCR products of *F. culmorum*. The SPR biosensor was applied to the detection of *F. culmorum* in fungal cultures and in naturally infected wheat samples.

Key words: DNA biosensor, Fusarium culmorum, PCR, Surface plasmon resonance, Wheat

1. Introduction

The conventional approach to detect pathogenic *Fusarium* species in infected plants and grains involves the isolation of the fungus into axenic culture or culturing on selective media followed by identification based upon morphological characteristics (1). Recently, more reliable molecular methods for the detection of several *Fusarium* species have been reported based on PCR amplification of specific DNA regions (2–4). Measurement of PCR products obtained by conventional PCR methods is performed by gel electrophoresis that allows discrimination of different DNA fragments by their molecular weight. Despite these methods being simple and relatively inexpensive, they use mutagenic or potentially mutagenic labeling reagents. Moreover, PCR could provide byproducts arising from the amplification of nonspecific sequences

Louise O'Connor and Barry Glynn (eds.), *Fungal Diagnostics: Methods and Protocols*, Methods in Molecular Biology, vol. 968, DOI 10.1007/978-1-62703-257-5_12,

with the same length of the desired fragment, leading to false-positive results. Therefore a hybridization assay is strongly recommended to confirm the selectivity of the reaction.

Surface plasmon resonance (SPR)-based DNA biosensors have been shown to be rapid, label-free, and selective tools for the detection of genetically modified organisms (GMOs) and of chemical and biological species, including fungal pathogens (5–9). By immobilizing a biological recognition molecule (probe) on the sensing surface, SPR biosensors provide a tool for real-time biospecific interaction analysis. The choice of a specific single-stranded DNA ligand allows rapid detection of the target DNA by hybridization with higher selectivity with respect to gel-electrophoresis analysis.

In this study, an SPR genosensor combined with PCR amplification has been used for the detection of *Fusarium culmorum* in pure fungal cultures as well as in wheat samples. *Fusarium culmorum* is a common pathogen of wheat involved in the etiology of Fusarium head blight (FHB), a worldwide disease causing total or partial ear premature senescence with consequent reduction of both crop yields and grain quality. In addition, *F. culmorum* can produce toxic secondary metabolites, mainly deoxynivalenol and zeralenone, which are mycotoxins harmful to humans and livestock (10–12).

2. Materials

Prepare all solutions using ultrapure water (e.g., produced by a Milli-Q system) and analytical grade reagents. Prepare and store all reagents at room temperature (unless indicated otherwise). Follow all waste disposal regulations when disposing waste materials.

2.1. Fungal Culture

1. Potato-dextrose-agar (PDA) (see Note 1).
2. Wickerham medium: 40 g glucose, 5 g peptone, 3 g yeast extract, 3 g malt extract, and water up to 1 L (see Note 2).

2.2. DNA Extraction from Fungal Cultures

1. Extraction buffer (SDS buffer): 200 mM Tris–HCl pH 7.8, 20 mM EDTA, 150 mM NaCl, 2% SDS, 1% β-mercaptoethanol, and 100 μg/mL proteinase K in water (see Note 3).
2. Phenol/chloroform/isoamylic alcohol (25:24:1, v/v/v).
3. Chloroform/isoamyl alcohol (24:1, v/v).
4. 3 M Sodium acetate, pH 5.2.
5. Absolute ethanol (ice cold).
6. 70% Ethanol (ice cold).

2.3. DNA Extraction from Wheat

1. Extraction buffer (CTAB buffer): 22 mM cetyltrimethylammonium bromide, 34 mM sarkosyl, 137 mM sorbitol, 22 mM EDTA, 1% cross-linked polyvinylpyrrolidone, 1.5 M NaCl in water (see Note 4).
2. Chloroform:isoamyl alcohol (24:1, v/v).
3. 5 M Potassium acetate.
4. Absolute ethanol (ice cold).
5. 70% Ethanol (ice cold).

2.4. DNA Quality and Quantification

1. DNA standard markers
2. 100 kb DNA ladder
3. 1× TAE buffer (Tris-Acetate-EDTA): 40 mM Tris, 20 mM acetic acid, and 1 mM EDTA (see Note **5**).
4. 1× TBE buffer (Tris-Borate-EDTA): 89 mM Tris, 89 mM boric acid, 2 mM EDTA (see Note **6**).
5. 0.7% Agarose gel.
6. 2% Agarose gel.
7. Ethidium bromide solution (see Note 7).

2.5. PCR Amplification

1. Specific primers: 5′-ATG GTG AAC TCG TCG TGG C-3′ (Fc01F) and 5′-CCC TTC TTA CGC CAA TCT CG-3′ (Fc01R) (see ref. (13)).
2. Deoxynucleotide triphosphates (dNTPs).
3. Hotmaster™ Taq DNA polymerase.
4. Taq buffer with Mg^{2+}.
5. PCR centrifugal filter devices.

2.6. SPR Probes

1. 25-mer target probe (Fc01P): 5′-CTT TAC CCC TCT GTT ACT AAA CTA T-3′.
2. 25-mer reference probe (Ref01P): 5′-ATC TAA AAT GCT TCC AAA CAG CGG C-3′.
3. 5’ biotin-modified target probe (Fc01P) and reference probe (Ref01P).

2.7. SPR Analysis

1. Biacore® X apparatus (Biacore International AB, Uppsala, Sweden).
2. Biacore SA sensor chips (carboxymethylated dextran matrix pre-immobilized with streptavidin).
3. Running buffer (PBS-E300 buffer solution): 20 mM Na_2HPO_4 (pH 7.4), 0.1 mM EDTA, 300 mM NaCl in water.
4. Conditioning solution: 1 M NaCl and 50 mM NaOH in water.
5. Sensor surface regeneration solutions: from 2 to 50 mM NaOH.

3. Methods

3.1. Preparation of Fungal Cultures

1. Culture fungal strains at 25°C on potato-dextrose-agar (PDA) (see Note 8).
2. Take mycelium plugs from 7-day-old colonies on PDA. Inoculate mycelium into 100 mL of Whickeram liquid media and incubate at 25°C for 48 h on a rotary shaker (110 rpm) (see Note 9).
3. Collect the mycelium by filtration through Whatman No. 1 filter paper and lyophilize (see Note 10).

3.2. DNA Extraction from Fungal Cultures

Genomic DNA from fungal cultures is extracted according to the protocol described by Sambrook et al. (14).

1. Grind finely lyophilized mycelium (30 mg) and add 600 μL of hot SDS extraction buffer (see Note 11).
2. After incubation for 90 min at 65°C (see Note 12), add 700 μL of phenol/chloroform/isoamyl alcohol (25:24:1, v/v/v) solution. Mix thoroughly by inversion to form a complete emulsion. Centrifuge at 10,000 × *g* for 20 min at 4°C (see Note 13).
3. Transfer the supernatant solution from the aqueous phase (top) into a new microcentrifuge tube. Add 700 μL of chloroform/isoamyl alcohol (24:1, v/v) and perform the above procedure (step 2) (see Note 13).
4. Transfer the supernatant solution containing the DNA from the aqueous phase (top) into a new microcentrifuge tube. Add 0.1 volumes of 3 M sodium acetate at pH 5.2 and 2.5 volumes of cold absolute ethanol. Store mixture solution at −20°C for 2 h (see Note 14).
5. Centrifuge at 12,000 × *g* for 20 min at 4°C to allow DNA precipitation. Discard supernatant.
6. Wash the pellet with 1 mL of 70% cold ethanol and centrifuge at 12,000 × *g* for 10 min at 4°C. Discard supernatant. Repeat the washing with 1 mL of 70% cold ethanol, centrifuge and discard supernatant.
7. Dry the precipitated DNA under vacuum (see Note 15). Rehydrate the pellet with 100 μL sterile water and leave overnight at 4°C to allow complete dissolving.

3.3. DNA Extraction from Wheat

Genomic DNA from wheat grain samples is extracted according to the method described by Wilson et al. (15), with minor modifications.

1. Grind finely wheat kernels. Add 20 mL of hot CTAB buffer to 4 g of milled wheat into a 50 mL plastic conical tube (see Note 16).

2. Incubate at 65°C for 1 h with frequently vigorous shaking.
3. Add 6.7 mL (one-third volume) of 5 M potassium acetate along with 6.7 mL (one-third volume) of chloroform:isoamyl alcohol (24:1). Mix vigorously and store at −20°C for 20 min.
4. Centrifuge the mixture at 1,900 × *g* for 15 min.
5. Transfer 600 μL of supernatant solution containing the DNA from the aqueous phase (top) into a new microcentrifuge tube.
6. Add 1.2 mL of cold absolute ethanol. Mix and store at −20°C for 1 h (see Note 14).
7. Centrifuge at 850 × *g* for 10 min at 4°C to allow DNA precipitation. Discard supernatant.
8. Wash the pellet with 1 mL of 70% cold ethanol and centrifuge at 850 × *g* for 5 min at 4°C. Discard supernatant. Repeat the washing with 1 mL of 70% cold ethanol, centrifuge and discard supernatant.
9. Dry the precipitated DNA under vacuum (see Note 15). Rehydrate the pellet with 300 μL sterile water and leave overnight at 4°C to allow complete dissolving.

3.4. DNA Quality and Quantification

1. Prepare the 0.7% agarose gel by mixing 0.7 g agarose and 100 mL 1× TAE buffer, and then adding 2 μL of ethidium bromide solution (10 mg/mL). Stir to mix (see Notes 7 and 17).
2. Load 5 μL DNA standard markers and 10 μL DNA samples into separate wells.
3. Run the gel for 20–30 min at 60 mA.
4. Compare with DNA standard markers (see Note 18).

3.5. PCR Amplification

1. In a microcentrifuge tube prepare a PCR mixture (50 μL) containing: 1 ng fungal DNA or 30 ng of total DNA extracted from wheat samples, 100 μM of each deoxynucleotide triphosphate (dNTP), 100 nM of each oligonucleotide primer, and 2 U of Hotmaster™ Taq DNA polymerase in Hotmaster™ Taq buffer with Mg^{2+}.
2. Perform amplifications using following conditions: denaturation at 94°C for 2 min, then 40 cycles of 94°C for 50 s, 57°C for 50 s, and 72°C for 1 min, and a final extension at 72°C for 7 min followed by cooling to 4°C until recovery of samples (see Note 19).
3. In order to allow concentration and removal of primers and unincorporated nucleotides, purify PCR products by using Montage™ PCR centrifugal filter devices prior to SPR analysis according to the manufacturer's protocol (see Note 20).

3.6. PCR Product Quality

1. Prepare the 2% agarose gel by mixing 2 g agarose and 100 mL TAE buffer, and then adding 2 μL of ethidium bromide (10 mg/mL). Stir to mix (see Notes 7 and 17).
2. Screen PCR products by gel electrophoresis on 2% agarose gel, stained with ethidium bromide. Load 5 μL DNA ladder and 10 μL DNA samples into separate wells. Run the gel for 20–30 min at 60 mA.
3. Visualize using an UV transilluminator and compare with a 100 kb DNA ladder (see Note 18).

3.7. SPR Analysis (see Note 21)

1. Immobilize the biotinylated probes by injecting 100 μL of a 1 μM probe solution in PBS-E300 (flow rate of 5 μL/min), by means of biotin–streptavidin interaction (see Note 22).
2. Add 10 μL formamide to 40 μL of purified PCR products and then adjust the total volume to 100 μL with 2× PBS-E300 buffer (see Note 23).
3. Prior to SPR analysis, subject the mixture containing the PCR product to denaturation at 95°C for 5 min followed by 1 min cooling on ice.
4. Inject 50 μL of PCR denatured product into the SPR apparatus. All SPR measurements are carried out at 25°C. The flow rate is 5 μL/min and the injection time is 10 min (see Note 24).
5. Regenerate the sensor surface by performing short injections (up to 1 min) of 2 mM NaOH to 50 mM NaOH until complete recovery of the baseline RU value on both flow cells (see Note 25).

4. Notes

1. Suspend 39 g of powder in 1 L of distilled water. Heat until completely dissolved and sterilize in the autoclave at 121°C for 15 min.
2. Weight 40 g glucose, 5 g peptone, 3 g yeast extract, 3 g malt extract and transfer to 1-L graduated cylinder or a glass beaker. Add 900 mL distilled water and mix. Make up to 1-L with distilled water and mix. Transfer the liquid media in a flask and sterilize it in the autoclave at 121°C for 15 min.
3. For preparing 10 mL SDS extraction buffer add 2 mL of 1 M Tris–HCl pH 7.8, 400 μL of 1 M EDTA, 300 μL of 5 M NaCl, 2 mL of 10% SDS, and make up to 9.85 mL with distilled water. Mix gently. Sterilize in the autoclave at 121°C for 15 min. Prior to start DNA extraction, add 100 μL of β-mercaptoethanol and 50 μL of 20 mg/mL proteinase K to the buffer solution.

4. For preparing 100 mL CTAB buffer add 2.0 g cetyltrimethylammonium bromide, 34 mM sarkosyl, 137 mM sorbitol, 4 mL 0.5 M EDTA, 28 mL 5 M NaCl, 40 mL water and mix gently. Make up to 100 mL with distilled water. Sterilize in the autoclave at 121°C for 15 min. Prior to start DNA extraction, add 1 g polyvinylpyrrolidone to the buffer solution.
5. TAE buffer is commonly prepared as a 50× stock solution. A 50× stock solution can be prepared by dissolving 242 g Tris in water, adding 57.1 mL glacial acetic acid, and 100 mL of 500 mM EDTA (pH 8.0) solution, and bringing the final volume up to 1 l. Sterilize in the autoclave at 121°C for 15 min. This stock solution can be diluted 50:1 with water to make a 1× working solution. This 1× TAE buffer contain 40 mM Tris, 20 mM acetic acid, and 1 mM EDTA.
6. TBE buffer is commonly prepared as a 10× stock solution. A 10× stock solution can be prepared by dissolving 1 g NaOH, 108 g Tris, 55 g boric acid, 7.4 g EDTA and bringing the final volume up to 1 l. Sterilize in the autoclave at 121°C for 15 min. This stock solution can be diluted 10:1 with water to make a 1× working solution. This 1× TBE buffer contain 89 mM Tris, 89 mM boric acid, 2 mM EDTA.
7. Ethidium bromide may be a mutagen, carcinogen, or teratogen agent. Material should be handled with care. Preparation of solutions of ethidium bromide and any operations should be conducted in a fume hood to prevent inhalation. Gloves, lab coat, and eye protection should be worn at all times. When UV source is used in work with ethidium bromide, avoid exposing unprotected skin and eyes to UV source.
8. Fungal strains can be obtained from the ITEM collection (CNR Institute of Sciences of Food Production, Bari, Italy, http://www.ispa.cnr.it/Collection/). In our study, fungal strain included: 20 strains of *Fusarium culmorum* (ITEM 627, 628, 5991, 5996, 6221, 6229, 6234, 6249, 6255, 6260, 6264, 6267, 6268, 6269, 6270, 6625, 6627, 6628, 6629, 6630), 4 strains of *Fusarium graminearum* (ITEM 633, 4432, 4433, 4440), 2 strains of *Fusarium proliferatum* (ITEM 1725, 2620), and 1 strain of *Fusarium avenaceum* (ITEM 3409), *Fusarium poae* (ITEM 3725), *Fusarium cerealis* (ITEM 666), *Fusarium verticillioides* (ITEM 3993), *Fusarium subglutinans* (ITEM 3922), *Fusarium sambucinum* (ITEM 846), *Fusarium acuminatum* (ITEM 797), *Fusarium oxysporum* (ITEM 3867), *Aspergillus ochraceus* (ITEM 4537), *Aspergillus carbonarius* (ITEM 4864), *Aspergillus niger* (ITEM 4736) and *Penicillium expansum* (ITEM 7011). The strains are transferred from tube of fresh cultures stored at 4°C using a sterile loop on the PDA plate with a one point inoculum for the *Fusarium* species and a 3 points inoculum for the *Aspergillus* and *Penicillium* species.

The inoculums is made using sterile condition under a sterile laminar flow hood and sterilizing the loop under the flame.

9. Inoculum is made on sterile condition by inoculating the flask by mycelia plugs using a wire sterile loop. A 100 mL of Wickerham medium is put in 250 mL flask and sterilized at 121°C for 15′. For the sterility of the inoculum it is important to flame the neck and the rim of the flask before and after the mycelium inoculum. After 48 h incubation and before fungal culture filtration, in order to avoid bacteria contaminations it is important to check the purity of the culture by observing it under a light microscope.
10. Filtrate the culture using a Büchner funnel, a vacuum flask connected to a vacuum pump and a Whatman filter paper. After the liquid filtrated is completely passed through the paper, recover the mycelium in a sterile falcon using a spatula and store at −20°C.
11. Lyophilized mycelium is finely ground with a mortar and pestle. An aliquot is weighted in a microcentrifuge tube where hot (65°C) SDS buffer is added.
12. Place the microcentrifuge tube into a 65°C water bath for 90 min. Shake frequently the microcentrifuge tube by vigorous shaking.
13. Two phases separated by a white layer will be observed.
14. Carry out this step quickly to avoid heating of ethanol. An ice bath can be used during this operation.
15. SpeedVac® System has been found useful for this operation. Drying can be carried out also by a desiccator for 20–30 min. Do not allow the DNA to over dry or it will be hard to redissolve.
16. The Cyclotec™ Sample Mill with a 0.5-mm mesh screen was found suitable. Hot (65°C) CTAB buffer must be used.
17. Add agarose to 1× TAE in a 250 mL conical flask, stir vigorously to mix and heat the mixture in a microwave oven for approximately 2 min until the gel is completely melted, paying attention to avoid boiling. Allow to cool for approximately 5 min. Then add ethidium bromide and mix. Pour the gel slowly into the gel tray avoiding bubbles. Carefully place the comb. Allow the gel to set for at least 30 min at room temperature. Then place it in the gel tank and submerge with 1× TBE buffer to 2–5 mm depth.
18. In our study, gels were visualized under UV light on a ChemiDoc system (Bio-Rad Laboratories S.r.l., Segrate, Italy) and analyzed using a Bio-Rad Quantity One software. Presence of a highly resolved band indicates good quality DNA. Presence of smeared band indicates DNA degradation.

19. Tubes without DNA template were included in each experiment and used as blank control samples.
20. For *Fusarium culmorum*, the PCR amplified product should be 0.57 kb.
21. In our study, SPR analyses were performed by a Biacore® X apparatus (Biacore International AB, Uppsala, Sweden), using Biacore SA sensor chips (carboxymethylated dextran matrix pre-immobilized with streptavidin) for all experiments. The presence of two flow cells on the sensor chip allowed the immobilization of Fc01P on flow cell 1 (FC1), as probe for target DNA detection, and Ref01P on flow cell 2 (FC2), as a reference probe intended to investigate nonspecific interactions. PBS-E300 buffer solution was used as running buffer for immobilization of probes on sensor chips and for hybridization experiments. Running buffer should be filtered through 0.2 μm cellulose filter and daily degassed for 30 min under vacuum and stirring. In order to avoid bubble formation in the Biacore X microfluidic system, a cleaning step is weekly recommended by flowing 0.5% (w/v) SDS in water (BIA maintenance kit) for 22 min followed by 50 mM Glicyne–NaOH, pH 9.5 (BIA maintenance kit) for 3 min and running buffer for 10 min.
22. Prior to probe immobilization, condition sensor chips with three consecutive 1-min injections of a solution containing 1 M NaCl and 50 mM NaOH. In our experiments, the immobilization procedure based on streptavidin–biotin binding, resulted in signal shifts of 1,622 ± 35 RU (n = 7) on flow cell 1 (FC1) for Fc01P probe (target DNA detection) and 1,568 ± 64 RU (n = 7) on flow cell 2 (FC2) for Ref01P probe (reference).
23. In order to establish optimum hybridization conditions, PCR product solutions containing 20 mM Na_2HPO_4 (pH 7.4), 0.1 mM EDTA, NaCl at concentration from 0 to 600 mM, and denaturing agents (urea or formamide) at concentration from 0 to 30% were subjected to SPR analysis. The use of 10% formamide and 300 mM NaCl in the testing solution allowed a nearly fourfold increase of the hybridization efficiency with respect to reference conditions (i.e., PBS-E150 buffer solution: 20 mM Na_2HPO_4 [pH 7.4], 0.1 mM EDTA, 150 mM NaCl).
24. A real-time baseline correction was automatically performed during each measurement by subtracting the reference signal recorded in flow cell 2 (FC2) from the target signal recorded in flow cell 1 (FC1). The analytical signal was represented by the resonance unit (RU) shift measured during sample injection, i.e. the difference between the RU value recorded after washing the sensor surface following hybridization reaction

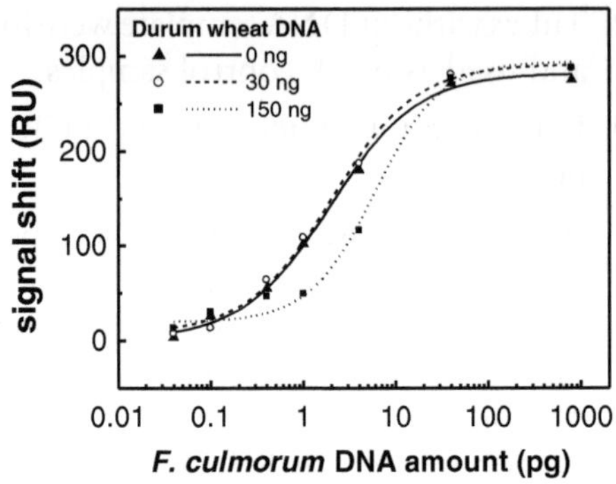

Fig. 1. Correlation between RU shift and the amount of starting *Fusarium culmorum* genomic DNA template (ITEM 627) in PCR, at different concentration of durum wheat DNA. All values were recorded in optimum experimental conditions (20 mM Na_2HPO_4 [pH 7.4], 0.1 mM EDTA, 300 mM NaCl, 10% formamide). Reproduced from ref. (7) with permission from Elsevier.

and the baseline RU value. In order to evaluate the matrix effect that could affect the SPR analysis of *F. culmorum* in wheat samples, PCR amplifications of different amount of genomic DNA of *F. culmorum* ranging from 0 to 800 pg were performed in presence of different amounts (0, 30, or 150 ng) of DNA of durum wheat buds grown under sterile conditions, and the relevant PCR products were detected by SPR biosensor (Fig. 1). Reliable results (similar to those obtained with pure culture) were observed only for sample with 30 ng durum wheat DNA. In the presence of 150 ng durum wheat DNA, RU shifts were significantly lower with respect to the corresponding values recorded with pure *F. culmorum* DNA, indicating an inhibitory effect of the matrix on PCR amplification. Detection limits of 0.05 pg (corresponding to a signal shift of 10.1 RU) and 0.06 pg (corresponding to a signal shift without correction for blank of 17.4 RU) of *F. culmorum* DNA template were observed in the absence and presence of 30 ng of durum wheat DNA, respectively.

25. The sensor surface is regenerated under strong alkaline conditions (pH > 11) allowing at least 75 hybridization cycles with a sensitivity loss lesser than 15%. NaOH at higher concentrations is used for sensor surface regeneration after a large number of hybridization cycles.

Acknowledgments

This work was supported by the Italian Ministry of Education, University and Research (MIUR), Project no. 12792 "SINSIAF".

References

1. Nirenberg HI (1981) A simplified method for identifying Fusarium spp. occurring on wheat. Can J Botany 59:1599–1609
2. Edwards SG, Pirgozliev SR, Hare MC, Jenkinson P (2001) Quantification of trichothecene-producing Fusarium species in harvested grain by competitive PCR to determine efficacies of fungicides against fusarium head blight of winter wheat. Appl Environ Microbiol 67:1575–1580
3. Yli-Mattila T, Paavanen-Huhtala S, Jestoi M, Parikka P, Hietanieme V, Gagkaeva T, Sarlin T, Haikara A, Laaksonen S, Rizzo A (2008) Real-time PCR detection and quantification of Fusarium poae, F. graminearum, F. sporotrichioides and F. langsethiae in cereal grains in Finland and Russia. Arch Phytopathol Plant Protection 41(4):243–260
4. Brunner K, Mach RL (2010) Quantitative detection of fungi by molecular methods: a case study on Fusarium. In: Voigt K, Gherbawy Y (eds) Molecular identification of fungi. Springer, Berlin, Heidelberg, pp 93–106
5. Homola J (2008) Surface plasmon resonance sensors for detection of chemical and biological species. Chem Rev 108:462–493
6. Homola J (2006) Surface plasmon resonance based sensors. Springer, Berlin, Heidelberg, p 251
7. Zezza F, Pascale M, Mulè G, Visconti A (2006) Detection of Fusarium culmorum in wheat by a surface plasmon resonance-based DNA sensor. J Microbiol Methods 66:529–537
8. Skottrup P, Hearty S, Frøkiær H, Leonard P, Hejgaard J, O'Kennedy R, Nicolaisen M, Justesen AF (2007) Detection of fungal spores using a generic surface plasmon resonance immunoassay. Biosens Bioelectron 22:2724–2729
9. Manera MG, Spadavecchia J, Leone A, Quaranta F, Rella R, Dell'Atti D, Minunni M, Mascini M, Siciliano P (2008) Surface plasmon resonance imaging technique for nucleic acid detection. Sens Actuators B Chem 130:82–87
10. Wagachaa JM, Muthomib JW (2007) Fusarium culmorum: infection process, mechanisms of mycotoxin production and their role in pathogenesis in wheat. Crop Prot 26:877–885
11. Parry DW, Jenkinson P, McLeod L (1995) Fusarium ear blight (scab) in small grain cereals-a review. Plant Pathol 44:207–238
12. Logrieco A, Bottalico A, Mulé G, Moretti A, Perrone G (2003) Epidemiology of toxigenic fungi and their associated mycotoxins for some Mediterranean crops. Eur J Plant Pathol 109:645–667
13. Nicholson P, Simpson DR, Weston G, Rezanoor HN, Lees AK, Parry DW, Joyce D (1998) Detection and quantification of Fusarium culmorum and Fusarium graminearum in cereals using PCR assays. Physiol Mol Plant Pathol 53:17–37
14. Sambrook J, Fritsch EF, Maniatis T (1989) Molecular cloning: a laboratory manual, 2nd edn. Cold Spring Harbor Laboratory Press, New York
15. Wilson A, Simpson D, Chandler E, Jennings P, Nicholson P (2004) Development of PCR assays for the detection and differentiation of Fusarium sporotrichioides and Fusarium langsethiae. FEMS Microbiol Lett 233:69–76

Chapter 13

Hyperbranching Rolling Circle Amplification, An Improved Protocol for Discriminating Between Closely Related Fungal Species

Jiufeng Sun and Sybren de Hoog

Abstract

Hyperbranching Rolling Circle Amplification (HRCA) is a technique derived from Rolling Circle Amplification (RCA) in which DNA polymerase replicates circularized oligonucleotide probes under isothermal conditions with either linear or geometric kinetics. Since its first introduction HRCA has been proven to be a robust DNA amplification technique for detecting pathogenic fungi and other applications, allowing rapid detection of nucleic-acid sequences with high specificity. Here we describe an improved protocol of HRCA, which is both specific and sensitive for detecting low copy numbers of template DNA. The test can be performed within one working day in routine molecular labs.

Key words: Rolling circle amplification, Isothermal polymerase, Padlock probe, Ligation, Single nucleotide polymorphism, Identification

1. Introduction

Nucleic-acid amplification technology has greatly increased our ability to clarify detailed questions about genotype or transcriptional phenotype in small biological samples, and has provided the impetus for many significant advances in biology. The methods have revolutionized fungal identification, which has moved from the specialist to the routine laboratory. To the wealth of available methodologies, the use of circularizable oligonucleotides, termed "padlock probes" initially used in the detection of repeated alphoid sequences in metaphase chromosomes of humans (1), has been added. The high sequence specificity of padlock probes allowing discrimination

Louise O'Connor and Barry Glynn (eds.), *Fungal Diagnostics: Methods and Protocols*, Methods in Molecular Biology, vol. 968, DOI 10.1007/978-1-62703-257-5_13, © Springer Science+Business Media New York 2013

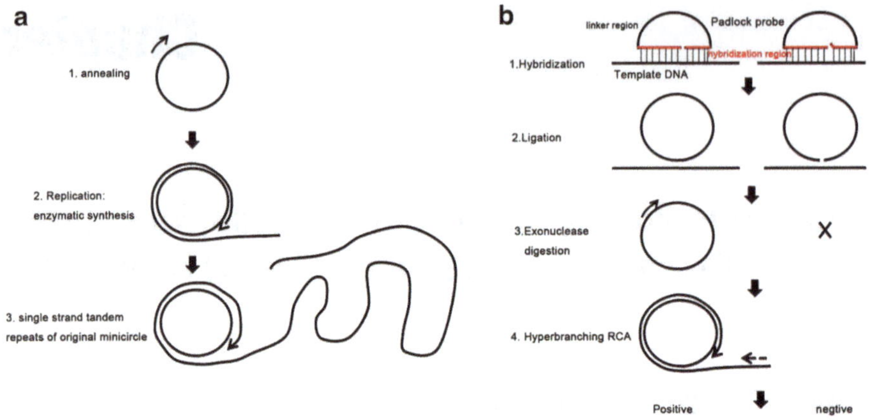

Fig. 1. Diagramatic representation of RCA (*Panel A*) and HRCA (*Panel B*) method. *Panel A*: The original RCA reaction can be divided in three major steps: (1) *mini-circle* denaturation and annealing of RCA primer, (2) Replication, (3) Single-strand tandem repeats of original *mini-circle*. During the first step the mini-circle is denatured followed by annealing of the RCA primer. Then, catalyzed by specific enzymes with strand replication activity (Bst DNA polymerase or Phi 29 DNA polymerase) the single-strand tandem replication reaction occurs. *Panel B*: The improved HRCA comprises four major steps: (1) DNA denaturation and hybridisation of Padlock probe, (2) Ligation reaction, (3) Exonuclease digestion, (4) Hyperbranching PCR reaction. In the first step the DNA is denatured and incubated overnight with Padlock probes which consist of two regions for RCA PCR primer annealing. The two ends of Padlock probe hybridize adjacent to each other on the target sequences. Only when the two ends of the probes are a perfect complement to the template DNA can hybridization occur followed by ligation during the ligation reaction (step 2). Unligated probes will be digested by Exonuclease (step 3) giving negative results. Ligated probes will be exponentially amplified during the subsequent PCR reaction (step 4) and resulting in long single-strand DNA being produced. Following this the RCA primer will be start another strand replication using the new single-strand DNA, resulting in different length double-strand DNA giving positive signals.

of point mutations in situ has prompted the generation of series of techniques such as HRCA suitable for the differentiation of molecular siblings that are increasingly being described (2).

As the original formulation of Hyperbranching Circle Amplification (HRCA) (3, 4), the Rolling Circle Amplification (RCA) reaction involves multiple rounds of DNA polymerase extending a circle-hybridized primer by continuously progressing around the circularized DNA probe isothermally (Fig. 1a). Several thousands of sequence-complementary tandem repeats of the original DNA mini-circle are yielded. RCA was normally used to generate line probes or repeat sequences of circled target sequence, while HRCA, a more complicated version of RCA, could be used to discriminate point mutation. During the HRCA reaction (Fig. 1b), specially designed padlock probes hybridize to target DNA or RNA, whereby the two ends of the probe become juxtaposed and are joined by a DNA ligase only when both ends of the probe show perfect complementarity. This permits the detection of single-nucleotide mismatches and prevents nonspecific amplification. In the presence of two primers, each hybridizing to one of the complementary strands, a complex pattern of DNA strand displacement generates over 10^9 copies of each circle in 90 min when catalyzed by a powerful isothermal DNA polymerase (2). To date, HRCA has mainly been used for the detection of bacteria (5) and viruses (6). In fungi, the technique has been applied to a number of genera, among which are *Cryptococcus*, *Trichophyton*, *Candida*, *Aspergillus*, *Scedosporium*, *Penicillium*, and *Fonsecaea* (7–12). The technique can be used for the detection of any species of fungi if the informative site (sequence heterogeneity) can be found. Here we introduce the workflow of HRCA with a practical protocol to be performed in the routine molecular laboratory.

2. Materials

Prepare all solutions using ultrapure water (MQ H_2O, 18 MΩ cm at 25°C) and analytical grade reagents. Prepare and store all reagents at room temperature (unless indicated otherwise). Follow all waste disposal regulations carefully.

2.1. DNA Extraction

TEx-buffer: 400 mM Tris, 10 mM Na–EDTA, pH 8.5-9.0.

Proteinase K: Dissolve 10 mg Proteinase K in 1 mL ultrapure water. Prepare a 1/200 dilution, aliquot and store at −20°C.

SEVAG: Chloroform:isoamyl-alcohol (24:1).

RNAse: Dissolve 10 mg pancreatic RNAse in 1 mL 0.01 M Na-acetate, pH 5.2. Heat the solution to 100°C for 15 min. Cool slowly to room temperature. Adjust pH by adding 1 M Tris, pH 7.4. Make a 1/100 dilution, aliquot and store at −20°C.

Tris-buffer: 10 mM Tris, pH 8.

TE-buffer: 10 mM Tris, 1 mM Na–EDTA, pH 8.

10% CTAB: 10 g cetyltrimethylammoniumbromide solved in 100 ml MQ H_2O.

Glass beads (Sigma)..

Bench top centrifuge.

NanoDrop DNA concentration detector (Thermo Scientific).

3. Pre-amplification

10× PCR buffer: 80 mM Tris-SO_4, 2 mM $MgSO_4$, 20 mM $(NH_4)_2SO_4$, 5% Glycerol, 5% DMSO, 0.06%. 0.05% Tween-20, pH 9.2 25°C.

dNTPs: 1.0 mM dNTPs each in 20 μL reaction system.

DNA polymerase: 5 U DNA polymerase (Takara, Japan) in a 50 μL reaction.

3.1. Ligation Reaction

10× Ligation buffer: 20 mM Tris–HCl (pH 7.5), 20 mM KCl, 10 mM $MgCl_2$, 0.1% Igepal, 0.01 mM rATP, 1 mM DTT.

Isothermal DNA ligase: 2 U *pfu* DNA ligase (Stratagene, USA) in each 10 μL reaction.

3.2. Exonucleolysis

10× Exonucleolysis buffer: 670 mM glycine–KOH, 67 mM $MgCl_2$, 100 mM 2-mercaptoethanol, pH 9.5, 25°C.

Exonuclease I and III: 10 U each of exonucleases I and III (New England Biolabs) in each 20 μL reaction volume.

3.3. RCA Reaction

10× RCA amplification reaction buffer: 20 mM Tris–HCl (pH 8.8), 10 mM KCl, 10 mM $(NH_4)_2SO_4$, 4 mM $MgSO_4$, 0.1% Triton X-100.

Bst DNA polymerase: 4 U *Bst* DNA large fragment polymerase (New England Biolabs) in each 20 μL reaction.

DNTPs: 1.0 mM dNTPs in each 20 μL reaction.

3.4. Amplicon Detection

50× TAE buffer: 2 M Tris–acetic acid, 100 mM EDTA, pH 8.5.

6× Loading buffer: 15% Ficoll-400, 66 mM EDTA, 17.8 mM Tris–HCl (pH 8.0, 25°C), 0.102% SDS, 0.09% bromophenol blue.

1% Agarose gel: 1 g agarose in 100 mL 1×TAE buffer.

Ethidium bromide stain: 10 mg/mL ethidium bromide, use 2 μL for 100 mL 1% agarose gel.

4. Methods

Carry out all procedures at room temperature unless specified otherwise.

4.1. DNA Extraction (see Note 1)

1. Transfer approximately 0.5 g mycelium of 14-day-old cultures are transferred to a 2 mL Eppendorf tube containing 400 μL TEx and glass beads.
2. Homogenize the fungal material using a vortex for 1 min.
3. Add 120 μL SDS 10% and 10 μL Proteinase K and incubate for 30 min at 55°C.
4. Vortex the mixture for 3 min.
5. Add 120 μL 5 M NaCl and 1/10 (V/V) CTAB 10% solution and incubate for 60 min at 55°C.
6. Transfer the mixture is then vortexed for 3 min.
7. Add 700 μL SEVAG and mix carefully by hand.

8. Centrifuge for 5 min at 4°C at 20,400 × *g*.
9. Transfer the supernatant to an Eppendorf tube with 225 μL 5 M NH_4-acetate.
10. Incubated for 30 min on ice water and centrifuge for 5 min at 4°C at 20,400 × *g*.
11. Transfer the supernatant to another Eppendorf tube with 0.55 vol isopropanol and centrifuge for 5 min at 20,400 × *g* force.
12. Wash the pellet with 1,000 μL ice-cold 70% ethanol.
13. Dry the pellet at room temperature and resuspend in 48.5 μL TE-buffer.
14. Determine the DNA concentration using the nano-drop DNA concentration detector at 260 nm.

4.2. Pre-amplification

For pre-amplification, universal primers may be used. In fungi for example ITS4 and ITS5 (Table 1) are used to amplify the Internal Transcribed Spacer region of rDNA (see Note 2). PCR conditions differ depending on the target gene. The program using primers ITS4 and ITS5 primers is as follows: 95°C for 5 min, followed by 35 cycles of 95°C for 30 s, 55°C for 30 s, and72°C for 1 min, with final extension at 72°C for 10 min. Amplification products are detected by electrophoresis on 1% agarose gels.

4.3. Padlock Probe Design

The Padlock probe is used as template in the HRCA reaction after a successful ligation reaction. The Padlock probe comprises two parts. One is a universal linker region, the other is the hybridization region which needs to be designed by incorporating the specific DNA sequence of the target species of interest. The first step is to gather as much possible reference DNA sequences of the target gene from both the target and closely related species (in this case the ITS1 DNA sequence in target species of *Penicillium marneffei* and closely species in the genus *Penicillium*). Then, DNA sequences of candidate gene of the target species and closely species are aligned using software such as free version of Clustal X1.81 (13). The informative nucleotide polymorphisms are identified within the alignment file (Fig. 2). Padlock probes targeting the selected regions should be designed with minimum secondary structure (excluding dimers, hairpins, and mismatches) (see Note 3). The linker regions of the padlock probe should be designed to minimize cross-reaction with closely species. Specificity of the probes is confirmed by BLAST analysis in GenBank. The Padlock probe and RCA1, RCA2 primers used in this method (Table 1) are from a previously published paper targeting *Penicillium marneffei* (11).

Table 1
RCA padlock probes and padlock probe-specific primers used in this chapter

Oligonucleotides	Sequences
ITS4	TCCTCCGCTTATTGATATGC
ITS5	GGAAGTAAAAGTCGTAACAAGG
RCA1	5′-CGCGCAGACACGATA-3′
RCA 2	5′-ATGGGCACCGAAGAAGCA-3′
PmPL1	5′-P-127AACGTCCCCCGGCGACCGG109 gatca**TGCTTC TTCGGTGCCCATT**ACGAGGTGCGGATAGCTA**CCGCGCAGACACGATA**gtcta 143CGCGGGCCCGGGGACA128-3′
PmPL2	5′-P-199GTACTCAGACAGTCCATCTT180 gatca**TGCTTCTTCGGTGCCCATT**ACGAGGTGCGGATAGCTA**CCGCGCAGACACGATA**gtcta 217TTTTGACAATTTTCATG200-3′
PmPL3	5′-P-421GGGACCACACCCAACACAC399 gatca**TGCTTCTTCGGTGCCCATT**ACGAGGTGCGGATAGCTA**CCGCGCAGACACGATA**gtcta 438TCGGGCAGGTCCCCGGA422-3′

The binding arms of the padlock probes which are joined by the backbone of the probe including the nonspecific linker region are underlined. The binding region of the RCA1 and RCA2 are in bold lettering. P: 5′ = phosphorylation

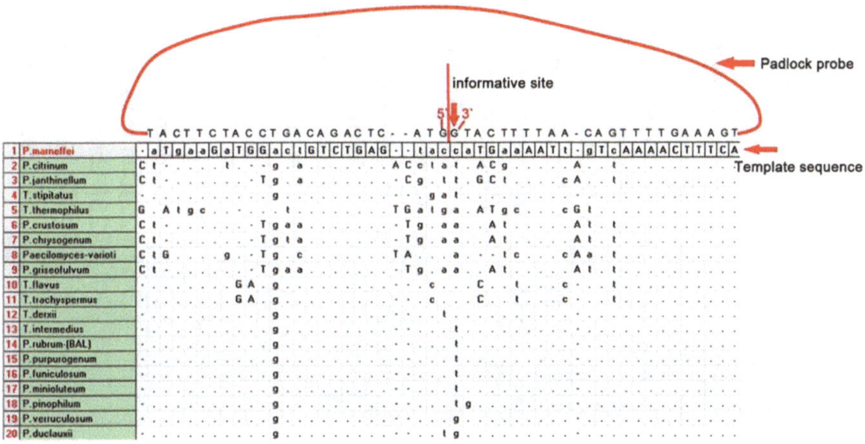

Fig. 2. Informative site detection in the ITS1 DNA sequence between *P.marneffei* and other closely fungi. The informative site, padlock probe and template DNA sequence are indicated with *arrow*. The linker region of Padlock probe is marked. The 5′ and 3′ end of Padlock probe are marked in figure.

4.4. Ligation of Padlock Probe

1. Combine 1 μL of amplified product (from Subheading 3.2) with 2 U *pfu* DNA ligase and 0.1 μM padlock probe in 10× ligation buffer final volume 10 μL.
2. Incubate 5 min at 94°C followed by 5 cycles at 94°C for 30 s and 4 min ligation at 63°C (see Note 4).

4.5. Exonucleolysis

1. Perform exonucleolysis (see Note 5) in a 20 μL reaction volume by adding 10 U of each exonuclease I and III to the ligation mixture (from Subheading 3.4).
2. Incubate at 37°C for 30 min, followed by 94°C for 3 min to inactivate the exonuclease.

4.6. HRCA Reaction

1. Perform RCA in 25 μL reaction volumes containing 0.25 μM each of RCA1 and RCA2, 4 U *Bst* DNA large fragment polymerase, 10× RCA amplification reaction buffer and 2 μL ligation product as template.
2. Incubate the reaction mixture at 65°C in a water bath for 60 min with final heating at 85°C for 2 min to terminate the reaction (see Note 6).

5. Data Analysis

1. Amplified products together with the Smart DNA ladder should be analyzed by electrophoresis on 1% agarose gels stained with ethidium bromide (see Note 7) and photographed under UV Light (see Note 8). Positive signals are obtained

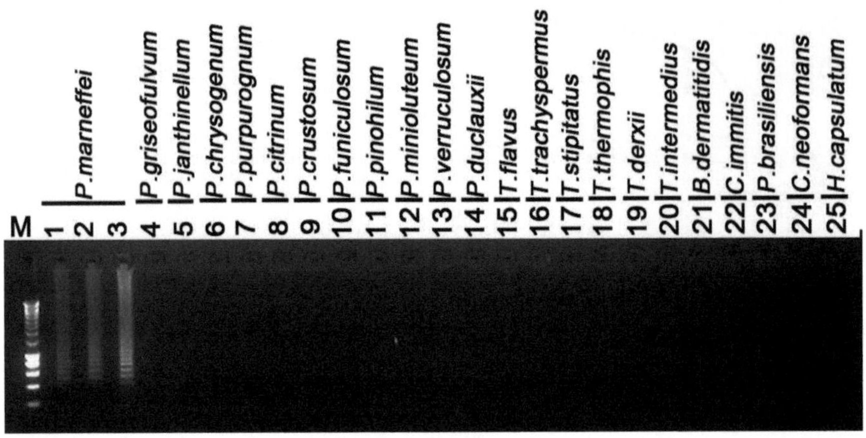

Fig. 3. RCA amplicons detected using electrophoresis on 1% agarose gel. *Lanes* 1–3 show positive samples giving a ladder pattern. *Lanes* 4–25 show no banding pattern indicating that the target sequence is not present.

when the designed padlock probes are a perfect complement to the targeted DNA sequence of *P.marneffei*, which are observed as a ladder-like pattern on a gel (Fig. 3 lane1-3). For closely related organisms there are no amplified products because the designed padlock probes are not a 100% match with the template DNA (Fig. 3 lanes 4–25). When the exonucleolysis step was omitted, unspecific bands appeared, but this did not interfere with the HRCA reaction (11).

6. Notes

1. DNA extraction protocols vary with the samples used. Here we showed the protocol frequently used for pure cultures of filamentous fungi.
2. The ITS region is widely used as target sequence for the identification of pathogenic fungi, applying universal primers ITS4 and ITS5. However, many fungi show insufficient diversity in ITS, and then hypervariable partial genes and introns are used, such as tubulin, actin, calmodulin, or translation elongation factor 1-α.
3. The T_m of the 5′ end probe binding arm is close to ligation temperature (63°C). To increase its discriminative specificity, the 3′ end binding arm was designed with a T_m 10–15°C below ligation temperature.
4. The ligation temperature may differ with each probe. A temperature gradient (58–65°C) amplification program was applied to optimize the annealing temperature using the optimized temperature, the padlock probes could hybridize with template DNA with the highest efficiency, resulting in the highest yield of HRCA products judged by the strongest signal on gel.

5. When the exonucleolysis step was omitted, unspecific background bands appeared, but this did not interfere with the RCA reaction.
6. This step is used to inactivate *Bst* DNA polymerase.
7. Can be replaced with other low-toxic compounds, such as SYBRgreen I dye.
8. DNA markers may be varied, depending on the size of the expected products.

Acknowledgments

This work was supported by The Federation of European Biochemical Societies (FEBS) Chinese European Visiting Fellowships (CEVF-00007), and was partly supported by the International Program of Project 985 at Sun Yat-Sen University.

References

1. Nilsson M, Malmgren H, Samiotaki M, Kwiatkowski M, Chowdhary BP et al (1994) Padlock probes: circularizing oligonucleotides for localized DNA detection. Science 265:2085–2088
2. Lizardi PM, Huang X, Zhu Z, Bray-Ward P, Thomas DC et al (1998) Mutation detection and single-molecule counting using isothermal rolling-circle amplification. Nat Genet 19:225–232
3. Fire A, Xu SQ (1995) Rolling replication of short DNA circles. Proc Natl Acad Sci U S A 92:4641–4645
4. Liu D, Daubendiek SL, Zillman MA, Ryan K, Kool ET (1996) Rolling circle DNA synthesis: small circular oligonucleotides as efficient templates for DNA polymerases. J Am Chem Soc 118:1587–1594
5. Tong Z, Kong F, Wang B, Zeng X, Gilbert GL (2007) A practical method for subtyping of Streptococcus agalactiae serotype III, of human origin, using rolling circle amplification. J Microbiol Methods 70:39–44
6. Schubert J, Habekuss A, Kazmaier K, Jeske H (2007) Surveying cereal-infecting geminiviruses in Germany–diagnostics and direct sequencing using rolling circle amplification. Virus Res 127:61–70
7. Kaocharoen S, Wang B, Tsui KM, Trilles L, Kong F et al (2008) Hyperbranched rolling circle amplification as a rapid and sensitive method for species identification within the Cryptococcus species complex. Electrophoresis 29:3183–3191
8. Kong F, Tong Z, Chen X, Sorrell T, Wang B et al (2008) Rapid identification and differentiation of Trichophyton species, based on sequence polymorphisms of the ribosomal internal transcribed spacer regions, by rolling-circle amplification. J Clin Microbiol 46:1192–1199
9. Zhou X, Kong F, Sorrell TC, Wang H, Duan Y et al (2008) Practical method for detection and identification of Candida, Aspergillus, and Scedosporium spp. by use of rolling-circle amplification. J Clin Microbiol 46:2423–2427
10. Lackner M, Najafzadeh MJ, Sun J, Lu Q, de Hoog GS (2012) Rapid identification of pseudallescheria and scedosporium strains by using rolling circle amplification. Appl Environ Microbiol 78:126–133
11. Sun J, Najafzadeh MJ, Zhang J, Vicente VA, Xi L et al (2011) Molecular identification of Penicillium marneffei using rolling circle amplification. Mycoses 54:e751–e759
12. Najafzadeh MJ, Sun J, Vicente VA, de Hoog GS (2011) Rapid identification of fungal pathogens by rolling circle amplification using Fonsecaea as a model. Mycoses 54:e577–e582
13. Thompson JD, Gibson TJ, Plewniak F, Jeanmougin F, Higgins DG (1997) The CLUSTAL_X windows interface: flexible strategies for multiple sequence alignment aided by quality analysis tools. Nucleic Acids Res 25:4876–4882

Chapter 14

Loop-Mediated Isothermal Amplification-Based Detection of *Fusarium graminearum*

Ludwig Niessen

Abstract

Molecular biological detection and quantification of fungal DNA targets today relies mainly on the application of the polymerase chain reaction (PCR). However, this well-recognized technique necessitates the use of highly purified DNA, in a well-equipped lab environment by trained personnel. The method has therefore considerable restrictions when it comes to on-site testing applications for phytopathogenic, mycotoxigenic, and medically relevant fungi and yeasts. As opposed to PCR, molecular biological detection of fungal DNA with isothermal amplification methods is performed at constant temperature. This makes thermal cycling superfluous and enables application of molecular detection assays on site. Moreover, detection of amplification product is done in-tube with no post-amplification manipulations necessary. Here we describe the use of loop-mediated isothermal amplification (LAMP) with indirect in-tube detection of DNA amplification as a rapid and robust method. Detection of *F. graminearum* in pure cultures and in cereal samples will be described as an example for the high potential which LAMP may hold for future developments in fungal detection assays.

Key words: LAMP, Isothermal amplification, *Fusarium graminearum*, Detection, Cereal, Single grain analysis

1. Introduction

Since its first publication by Mullis and Faloona (1), the polymerase chain reaction (PCR) with heat stable *taq*-polymerase (2) has become the gold standard technique for detection and identification of microorganisms, including molds and yeasts. A vast array of food spoiling and toxinogenic species can now be analyzed, both qualitatively and quantitatively, using this technique (3). However, application of PCR-based amplification methods is still a time con-

Louise O'Connor and Barry Glynn (eds.), *Fungal Diagnostics: Methods and Protocols*, Methods in Molecular Biology, vol. 968, DOI 10.1007/978-1-62703-257-5_14, © Springer Science+Business Media New York 2013

suming, expensive, and cumbersome job due to its requirement for specific lab equipment, skilled personnel, and agarose gel electrophoresis (AGE) needed for product separation and visualization. These factors foil the use of PCR-based detection systems in on-site applications, e.g., agriculture, food and feed production, public health and safety, counter terrorism. Real-time PCR using fluorescent intercalating dyes or fluorescently labeled oligonucleotide probes have made AGE dispensable but equipment and chemicals are even more expensive and demand for skilled staff is even higher.

Loop-mediated isothermal amplification (LAMP) of DNA was recently described by Notomi et al. (4). The method makes use of the large fragment of the *Bst* DNA polymerase from *Bacillus stearothermophilus*. The large fragment of the enzyme contains the 5'->3'polymerase activity but lacks 5'->3' exonuclease activity. *Bst* DNA polymerase large fragment displaces third strand DNA during primer initiated polymerization of new DNA leaving a new single-stranded matrix DNA for primer annealing and DNA polymerization. Since *Bst* DNA polymerase has a very high activity, vast amounts of high molecular weight DNA are produced within short time. The exceptionally high specificity of LAMP is due to the fact that a set of four primers with six binding sites must hybridize correctly to their target sequence before DNA biosynthesis occurs. A third pair of primers (loop primers) can be added optionally to the reaction in order to further amplify the amount of DNA produced during LAMP (5). Detection of amplified product has originally been done by AGE. However, scientists soon became aware that LAMP bears an exceptionally high risk of cross-contamination of samples by aerosolized product. In order to prevent cross-contamination, methods for in-tube detection of DNA amplification were developed. In-tube detection of amplified product has been achieved by direct staining of double-stranded DNA using intercalating dyes, e.g., SYBR green 1, SYTO-9 (6) or by precipitation of the LAMP product with a fluorescently labeled cationic polymer (7). Indirect in-tube detection of LAMP product was done using pyrophosphate, a specific by-product of enzymatic DNA synthesis. Detection was done by measuring the turbidity of LAMP reactions resulting from precipitation of a magnesium–pyrophosphate complex (8). Indirect calcein fluorescence has been used to detect DNA synthesis by removing the manganese quencher from the molecule through complexation with pyrophosphate during production of DNA (9). Turbidimetry has also been used in real-time applications of the LAMP reaction resulting in quantitative data (10).

Only a few applications of LAMP for the detection of fungal organisms have been described to date. Detection of *Paracoccidioides brasiliensis* was described by Endo et al. (11). Ohori et al.(12) applied the method to the detection of the dematiaceous fungus *Ochroconis gallopava*, an infective organism affecting the nervous system and respiratory tract of humans, birds, and cats. Tomlinson et al. (13) published a LAMP-based assay for the detection of the oomycete *Phytophthora ramorum*. A method for detection, identification, and quantification of *Brettanomyces/Dekkera* yeasts in the brewing process was described by Hayashi et al. (14). Gadkar and Rillig (15) set up a LAMP assay for detection and identification of arbuscular mycorrhizal fungi (AMF, e.g., *Glomus intraracices*) in carrot roots. Recently, LAMP-based diagnosis has been used for diagnosis of medically important fungi, i.e., *Fonsecaea* spp. (16), *Penicillium marneffei* (17), *Cryptococcus neoformans* (18), and pathogenic *Emericella* species (19). Tomlinson et al. (20) described a LAMP-based assay for on-site detection of *Botrytis cinerea* in cut flowers and vegetables and Niessen et al. (21) demonstrated the usefulness of LAMP for the specific detection of *Fusarium graminearum* in cereal samples. They detected the target directly from infected grains and from fungal cultures without previous DNA extraction.

F. graminearum produces mycotoxins such like deoxynivalenol (DON) and nivalenol (NIV) as well as the estrogen analogous compound zearalenone (ZON), all of which were found in infected cereals at high concentrations (22). Moreover, *F. graminearum* together with *F. culmorum* has been suspected to be involved in the induction of beer gushing, a most undesirable condition in which bottled beer foams heavily and spontaneously upon opening (23).

We describe here the development and application of a LAMP-based assay for the detection of *F. graminearum* in samples of purified DNA, pure cultures, and infected cereal samples. However, the method as described here can be used for the detection of any other fungi provided a set of specific LAMP primers is available. Also the method described here for primer design can be generalized in terms of using different sequence data.

2. Materials

All buffers and solutions are prepared with ultrapure deionized water. All buffers and solutions are filter sterilized and stored as described below. Separate sets of pipettes are used for manipulations involving DNA and for non-DNA manipulations. Disposable filter tips must be used for all pipetting. Plasticware used is certified to be free of human DNA, DNAse and RNAse. Preparation of master mix and steps involving DNA must be performed in separate rooms in order to prevent contamination with target DNA. All buffers and DNA must be free from EDTA. If sample DNA has been dissolved in TE buffer, it is necessary to re-extract the DNA and dissolve in ultrapure water before addition to LAMP reactions (see Note 1).

2.1. LAMP Master Mix

1. 10× LAMP buffer: 200 mM MOPS, 100 mM KCl, 100 mM $(NH_4)_2SO_4$, pH 8.8.

   To prepare 50 ml weigh 2.093 g MOPS, 0.373 g KCl, and 0.661 g $(NH_4)_2SO_4$ and add ultrapure water to 40 ml. Stir until dissolved. Adjust pH to 8.8 using 2 M NaOH. Add to 50 ml volumetric flask and add water to final volume. Filter sterilize through 0.2 μm filter cartridge into sterile 50 ml Falcon tube. Store at ambient temperature (see Note 2).

2. $MgCl_2$ solution: 200 mM $MgCl_2$.
   Weigh 10.17 g of $MgCl_2 \times 6H_2O$ for 50 ml 1 M $MgCl_2$ stock solution. Fill to 40 ml and stir until dissolved. Add to 50 ml volumetric flask and add water to final volume. Filter sterilize through 0.2 μm filter cartridge into sterile 50 ml Falcon tube. Store at ambient temperature. For preparation of working solution add 800 μl ultrapure H_2O to 200 μl of the $MgCl_2$ stock solution. Store at ambient temperature (see Note 3).

3. Nucleotide mix: 10 mM each G, A, T, C.
   Mix 1 ml of a 100 mM stock of dGTP, dATP, dTTP, dCTP, and add 6 ml of ultrapure water. Distribute aliquots of 500 μl to sterile 1.5 ml reaction tubes. Store at −20 °C. Store at 4 °C after thawing.

4. *Bst* dilution buffer: 50 mM KCl, 10 mM Tris–HCl (pH 7.5), 1 mM dithiothreitol, 0.1% Triton X-100, and 50% glycerol.
   Weigh 186.4 mg KCl, 78.8 mg Tris–HCl, 7.7 mg dithiothreitol, 50 mg Triton X-100 and fill to 40 ml with 50% (v/v) glycerol in water. Stir until dissolved and adjust to pH 7.5 with NaOH. Transfer to 50 ml volumetric flask and add 50% glycerol to final volume. Filter sterilize through 0.2 μm filter cartridge into sterile 50 ml Falcon tube. Store at −20 °C. Solution will stay liquid in freezer.

5. *Bst* DNA polymerase, large fragment (New England Biolabs): 8,000 U/ml working solution.
   Mix 70 μl of the enzyme solution as supplied (120,000 U/ml) with 930 μl Bst-dilution buffer. Distribute aliquots of 50 μl to sterile 500 μl reaction tubes. Store at −20°C.

6. LAMP primer mix: solution for 80 reactions.
   If primers are delivered in solution, check they do not contain EDTA and that concentrations are 50 μM. If primers are delivered freeze dried, add ultrapure sterile water to prepare a 50 μM solution (see data sheet from supplier). Mix 64 μl each of primers FIP-GaoA ID4 and BIP-GaoA ID4, 32 μl each of primers loopF-GaoA ID4 and loopB-GaoA ID4, 8 μl each of

Table 1
Primers used in the LAMP assay for the detection of *F. graminearum*

Primer name	Sequence 5'->3'	Reference
F3-gaoA ID4	AGG GAG TCT TCA GTT CCT GA	[21]
B3-gaoA ID4	GTG AGG GGG CTT TGG ATC	[21]
FIP-gaoA ID4	CGC AAG TGA CGG CCC AGT TGC TTC GAG CCT CAG CAC CTA	[21]
BIP-gaoA ID4	TGC AAC AAG GCC ATT GAT GGC CGT TGG CGC CAT AGA ATG T	[21]
LoopF-gaoA ID4	GTT GCG AGA AAT GGC GCT TCC G	[21]
LoopB-gaoA ID4	ACA AGG ATA CCT TTT GGC AC	[21]

F3-GaoA ID4 and B3-GaoA ID4 and mix well. Store at 4 °C. See Table 1 for primer sequences.

7. Calcein amplification detection reagent: 1.25 mM solution of manganese quenched calcein in 1× LAMP buffer in 40% glycerin.

   Weigh 418.6 mg MOPS, 74.6 mg KCl, 132.2 mg $(NH_4)_2SO_4$, and 211.3 mg $MnSO_4 \times 1H_2O$. Add ultrapure water to 40 ml and stir until completely dissolved. Adjust pH to 8.8 using 2 M NaOH. Transfer to 50 ml volumetric flask and add ultrapure water to final volume. Add 77.8 mg calcein and mix until dissolved. Mix solution 1:1 with 80% (v/v) glycerol and filter sterilize through 0.2 μm filter cartridge. Distribute aliquots of 1–1.5 ml reactions tubes. Store at −20 °C. Solution will stay liquid in freezer.

2.2. Other Components

1. Disposable plasticware: 0.2 μm filter cartridge; 200 μl PCR tubes, sterile filter pipette tips, 1.5 ml and 500 μl reaction tubes.
2. Two sets of pipettes including 200–1,000, 50–200, 20–100, 2–20, 1–10 μl.
3. Gradient thermocycler.
4. 366 nm UV lamp.
5. Digital camera.
6. Ultrapure water.

3. Methods

3.1. Design of LAMP Primers for F. graminearum (see Note 4)

1. Select nt 839–2638 of the DNA sequence of the galactose oxidase gene (*gaoA*) of *F. austroamericanum* NRRL 2903 from GenBank under accession no. M86819 and copy sequence text to a text editor.
2. Remove all formatting and numbers from the text using the search and replace function and save document as txt-file using DOS formatting option.
3. Open the PrimerExplorer V.4 software tool available on the Eiken Genome site provided by Eiken Chemical Co. Ltd. (Tokyo, Japan) using the following link: http://www.primerexplorer.jp/e/.
4. Follow instructions for designing regular primers: Give txt-file previously generated under step **2** as target sequence file. Browse your computer to find location of the file and use automatic judgment as parameter setting and press primer design button to upload target sequence.
5. Press generate button and wait until primer sets have been generated. Usually up to five sets will be generated.
6. Press display button and wait for separate window to open. Target sequence with primers generated will result.
7. Check boxes for all primer sets and press confirm button. A primer set confirmation page will open in separate window with the sets of primers generated.
8. Select primer set with value for "dimer(minimum)dG" as close to −2.0 as possible for further in silico analysis.
9. Use the BLAST-algorithm under http://www.blast.ncbi.nlm.nih.gov/ to search GenBank for matches with other sequences for each primer. Ideally, primers should show high score hits exclusively with the target gene in the target species. Low score hits with a variety of other species will occur regularly.
10. If BLAST analysis retrieves high score hits also with nontarget species, repeat step 9 with primer set next most closely to dG = −2.0.
11. Print out sequence of target gene and mark positions of the primers selected for LAMP.
12. Design loop primers to enhance amplification reaction. LoopF primer is positioned between the 3' end of the binding site for primer F2 and the 5' end of the binding site of primer F1c. LoopB primer is positioned between the 5'end of the binding site of primer B2 and the 3' end of the binding site of primer B1c. Make sure both loop primers have Tm of ±60 °C (see Notes 5, 6).

13. Order primers according to the sequence given on the primer confirmation page (see step 7). Make sure LoopF primer is reverse complement, whereas LoopB primer is according with the target sequence.

3.2. Preparation of LAMP Master Mix

1. Solution per 25 μl reaction:
   Mix 2.5 μl 10× LAMP buffer, 1 μl 200 mM $MgCl_2$, 3.5 μl dNTP mix, 2.6 μl 80× LAMP primer mix, 1 μl *Bst* DNA polymerase 8,000 U/ml, 1 μl Calcein amplification detection reagent, 11.4 μl ultrapure water. Distribute 23 μl of master mix per 200 μl PCR reaction tube with tightly fitting lid (see Note 7).

3.3. Assay Set Up: Temperature Optimization

1. Dilute purified genomic DNA of a *F. graminearum* strain obtained from a recognized culture collection to a concentration of 100 ng/μl in ultrapure sterile water (see Note 8).
2. Prepare master mix as described above (Subheading 3.2) for 13 reactions and add 2 μl per reaction of the DNA dilution prepared in step **1**. Distribute 25 μl each to 200 μl PCR reaction tubes and close lids tightly.
3. Program a gradient thermal cycler to result in a gradient of constant incubation temperatures ranging from 60 to 70°C in 12 steps with 60 min incubation followed by heating to 95 °C for 2 min in order to stop the enzyme reaction. If no gradient thermal cycler is available, incubate at 65 °C for 60 min followed by incubation at 95 °C for 2 min. Repeat with increasing temperature until fluorescence signal gets weaker or until no more fluorescence signal occurs (see Note 9).
4. Put reaction tubes against a black background and inspect reactions for occurrence of bright green fluorescence under UV_{366nm} lamp in a dark room. Figure 1 shows LAMP reactions with *F. graminearum* DSM 4527 DNA incubated according to step 3 (see Note 10).
5. Select highest temperature at which a bright fluorescent signal is produced as incubation temperature for further experiments.

3.4. Assay Set Up: Sensitivity Testing

1. Prepare a tenfold serial dilution of reference DNA (see Subheading 3.3, step 1) in ultrapure sterile water.
2. Prepare master mix according as above and distribute 23 μl in 8 200 μl reaction tubes. Seal lids tightly.
3. Add 2 μl of each dilution prepared in step 1 to a separate reaction tube. Add 2 μl of ultrapure sterile water to tube no. 8 as negative control.
4. Incubate for 60 min at constant temperature according to step 5 under Subheading 3.3 followed by incubation at 95 °C for 2 min.

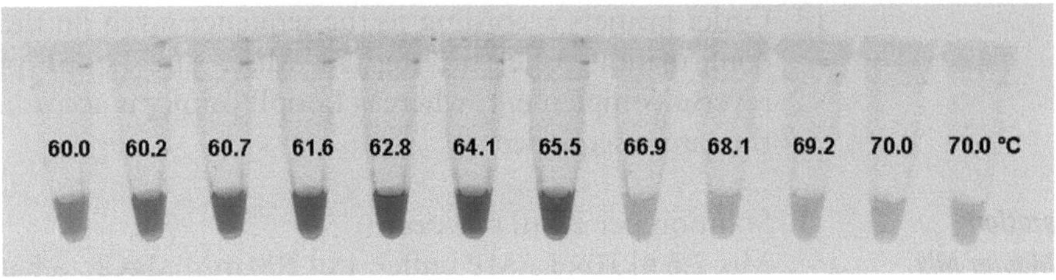

Fig. 1. Optimization of reaction temperature. LAMP reaction with primers GaoA ID4 and genomic DNA of *F. graminearum* DSM 4527 run for 60 min followed by inactivation at 95 °C for 2 min. Numbers indicate temperature under which the respective reactions were incubated. Bright fluorescence (= darker color of reactions) indicates in vitro DNA synthesis.

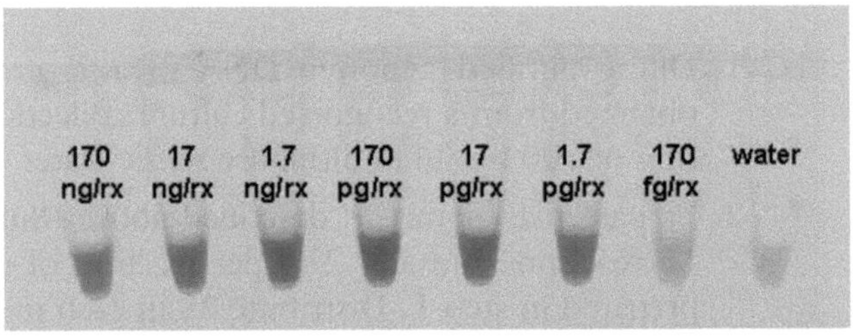

Fig. 2. Sensitivity testing of the LAMP assay for the detection of *F. graminearum*. LAMP reactions with primers GaoA ID4 using a serial dilution of DNA from *F. graminearum* NRRL 31084 as template after 60 min of incubation at 65 °C followed by inactivation at 95 °C for 2 min. Numbers above reactions indicate DNA concentration per reaction (rx). Water indicates negative sample with no DNA added.

5. Figure 2 shows LAMP reaction with primers specific for *F. graminearum* and a serial dilution of *F. graminearum* NRRL 31084 genomic DNA as target under UV_{366nm} light. Assay should detect DNA to a dilution of 10^{-6} or less.

3.5. Assay Set Up: Specificity Testing

1. Prepare purified genomic DNA of the most closely related species and dilute it to result in 100 ng/µl solutions in ultrapure sterile water. If assay is planned to be used in sample materials, prepare also genomic DNA from non-contaminated sample matrix or matrices (see Note 11).
2. Prepare a 100 pg/µl solution of the *F. graminearum* reference DNA in ultrapure sterile water as positive control.
3. Prepare sufficient Mastermix for the number of reactions to be run as described above (Subheading 3.2) and distribute 23 µl each to 200 µl PCR reaction tubes and seal tightly.
4. Add 2 µl of DNA solution to be tested per reaction. Add 2 µl of reference DNA to the positive control and 2 µl of ultrapure sterile water to the negative control and incubate for 60 min at constant temperature as described above (Subheading 3.3, step 5) followed by incubation at 95 °C for 2 min.
5. Inspection under UV_{366nm} light should result in a bright green fluorescence in the positive control only. If samples of test DNA result in a bright green fluorescence signal, the assay has to be adjusted for higher specificity.

3.6. Assay Set Up: Specificity Adjustment

1. Perform LAMP reaction with DNA samples of those non-target species which showed false-positive reaction under the conditions described above (Subheading 3.5) including the *F. graminearum* reference DNA and water as positive and negative controls, respectively. Increase incubation temperature in 0.5 °C steps until no false-positive reactions occur.
2. In situation where higher temperatures do not solve the problem, prepare separate master mixes and add 26 µl of the reference DNA and DNA of the species resulting in the strongest fluorescence signal in step 1. Prepare 8 200 µl PCR reaction tubes for each DNA to be tested and add from 0 to 1.75 µl of formamide in 0.25 µl steps (equal to 0–7% (v/v)). Add 25 µl of each reaction mixture into reaction tubes and seal tightly.
3. Incubate LAMP reactions at temperature 2 °C below the highest temperature used in step 1 for 60 min followed by incubation at 95 °C for 2 min.
4. Select conditions which avoid false-positive signal and repeat testing with samples used in step 1.
5. If there are still false-positive reactions, repeat with increasing incubation temperature until no false-positive reactions occur.
6. Continue testing with as many different species as reasonably possible in order to achieve an overview over the range of species detected by the assay.

3.7. Assay Set Up: Adjustment of Incubation Time

1. Prepare a serial dilution of the reference DNA in ultrapure sterile water and add to LAMP master mix according to Subheading 3.4, steps 1–3 (or optimized according to Subheading 3.6).
2. Remove samples from incubation every 10 min and cool the reaction mixture under cold tap water (see Note 12). Take photograph of reaction tubes under UV_{366nm} light before continuing incubation for another 10 min under the same conditions (see Note 13).
3. Repeat step 2 until an accumulated incubation time of 90 min is reached.
4. The accumulated time interval after which no more change in fluorescence occurs is the optimum incubation time. Typically this is between 35 and 50 min as shown in Fig. 3.

3.8. Assay Application: Direct Testing of Fungal Pure Cultures

1. Grow *F. graminearum* cultures on SNA (24) plates for 7–10 days at ambient temperature in diffuse day light (see Note 14).
2. Prepare LAMP master mix as described. Make sure to add an additional 2 µl of ultrapure sterile water per reaction to compensate for the sample DNA volume.

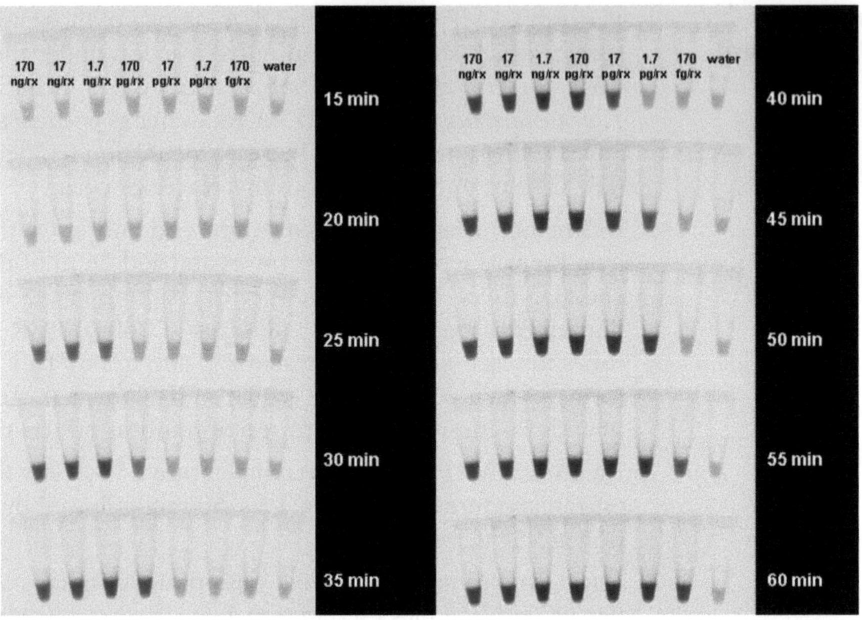

Fig. 3. Optimization of reaction time for the LAMP assay for detection of *F. graminearum*. A serial dilution of genomic DNA of F. graminearum NRRL 31084 was incubated under optimized conditions in LAMP master mix. Reactions were removed from incubation every 5 min and photographed under UV_{366nm} light. Single photos were mounted together after 60 min of cumulated reaction time. Numbers above reactions indicate DNA concentration per reaction (rx). Water indicates negative sample with no DNA added.

3. Add 25 μl of LAMP master mix per 200 μl PCR reaction tube and seal tightly.
4. Bend the tip of a steel inoculation needle to form a small hook. Make sure not to use a platinum needle since the material interferes with the LAMP reaction.
5. Flame sterilize needle and cool in agar plate before transferring a tiny yet visible portion of mycelium or spores into the LAMP master mix and detach material from needle by rotating it several times in the mix.
6. Remove needle and close reaction tube tightly before flame sterilizing again.
7. Repeat for each culture to be tested. Use purified reference DNA and a culture of the reference strain as positive controls. Use reaction without any fungal material as negative control.

3.9. Assay Application: Single Grain Analysis for Cereals

1. Prepare double amount of LAMP master mix per reaction as described (Subheading 3.4, step 1–3 (or as optimized under Subheading 3.6)) with the addition of 4 μl of ultrapure sterile water per reaction to compensate for the DNA sample volume and distribute 50 μl per reaction in 200 μl PCR reaction tubes and seal tightly (see Note 15).

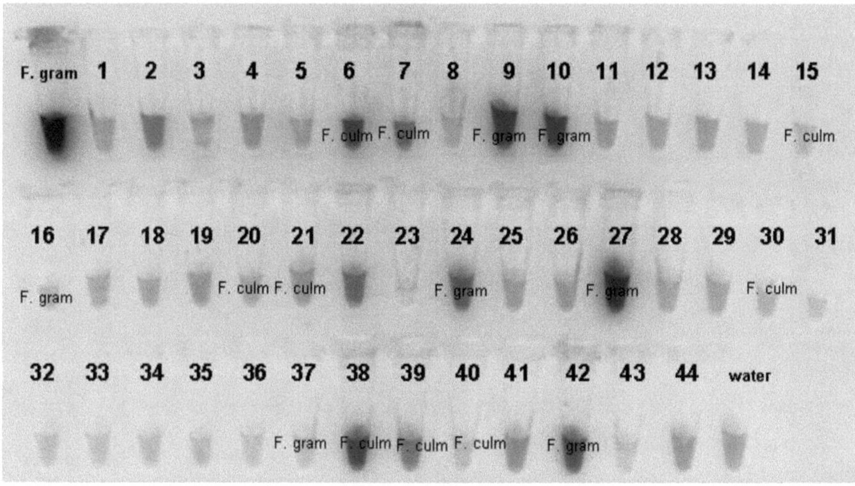

Fig. 4. Single grain analysis of barley grains with LAMP primers GaoA ID4. Grains were immersed in the LAMP master mix prior to incubation for 60 min at 65 °C and inactivation for 2 min at 95 °C. Grains were removed from the master mix and plated on SNA medium (24) prior to incubation for 10 days at ambient temperature under white light/UV light. Grains from which *F. graminearum* or *F. culmorum* were isolated are marked. Upper left reaction is the positive control. Lower right reaction is the negative control with water instead of DNA.

2. Insert one grain per reaction tube and reseal tightly. Vortex samples for 1 min.
3. Remove grains from LAMP master mix before incubation. Grains can be used for parallel microbiological analysis. Grains may be left in the master mix during incubation. However, results have been found to be less clear in terms of fluorescence visibility. However, grains once incubated cannot be used for microbiological analysis anymore.
4. Incubate LAMP reactions under optimized conditions.
5. Inspect for bright green fluorescence under UV_{366nm} light against a black background. Figure 4 shows results of single grain analysis of a barley sample with grains removed before incubation. Grains were analyzed for the presence of viable fungi using SNA medium.

3.10. Assay Application: Analysis of Bulk Grain Samples

1. Prepare three LAMP master mixes per sample plus controls as described above (Subheading 3.4, or as optimized in Subheading 3.6) and distribute 23 μl per 200 μl PCR tube. Seal reaction tubes tightly.
2. Measure 15 ml of grains from a bulk sample in a sterile Falcon tube and add sterile ultrapure water until grains are fully covered.
3. Vortex for 1 min at maximum speed or shake vigorously by hand for 1 min.

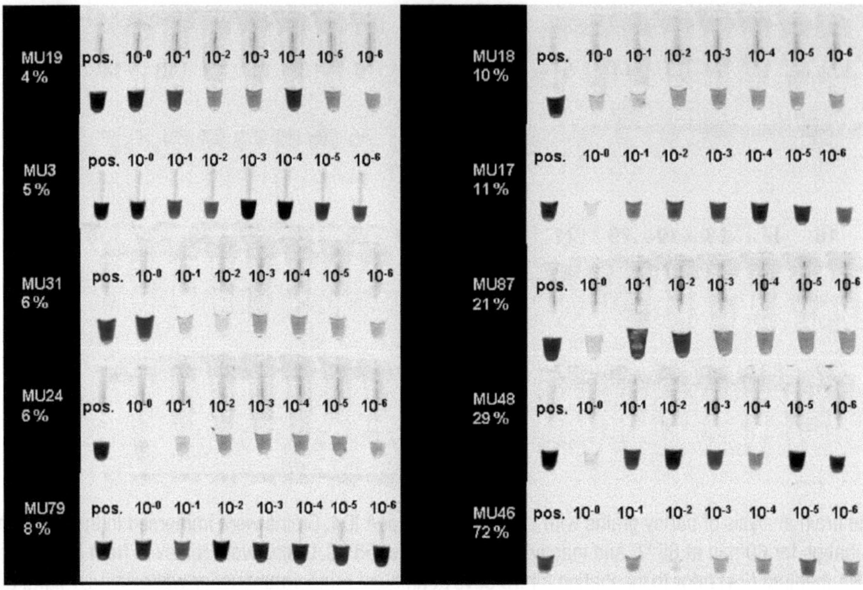

Fig. 5. Analysis of bulk barley grain samples with LAMP primer set tric ACL1 ID13 specific for *F. tricinctum* (27). Grains were washed with sterile ultrapure water. Undiluted washings and washings diluted tenfold in sterile ultrapure water were used as template in the LAMP reaction. pos. = positive controls using pure DNA of F. tricinctum. MU3–MU87 are sample numbers and the percentage of grains infected with F. tricinctum in respective samples.

4. Transfer 1 ml of the liquid to a fresh sterile 1.5 ml reaction tube and seal tightly.
5. Prepare five- or tenfold serial dilution of the washings for each sample in ultrapure sterile water as appropriate (here 10^{-1} to 10^{-6}).
6. Add 2 μl of the undiluted washing and 2 μl of each dilution per sample to master mix and seal tubes tightly.
7. Incubate under conditions optimized as described in Subheadings 3.6 and 3.7.
8. Inspect for bright green fluorescence under UV_{366nm} light against a black background. Figure 5 shows the results obtained with washings from samples of bulk barley grains and analysis for F. tricinctum (see Note 16).

4. Notes

1. EDTA is a highly active molecule and reacts with bivalent cations such as manganese and magnesium. EDTA will remove the manganese quencher from calcein and leads to false-positive reactions in the LAMP. As a general precaution to prevent heavy contamination of lab space with LAMP product DNA,

never open LAMP reaction tubes once the reaction has started. If reactions are separated on agarose gels make sure that transfer of sample-DNA is well separated from the place where master mixes are prepared. Use pipets which are never used to set up LAMP master mix.

2. Original descriptions of the10× LAMP buffer use Tris–HCl as the buffering substance. This compound is widely used in amplification reactions. However, according to our own experiments the reaction is considerably more sensitive if MOPS is used instead of Tris–HCl in the same concentration. The reaction also runs with HEPES as buffering component, but sensitivity was comparable to the sensitivity obtained with Tris–HCl.
3. The original 10× LAMP buffer provides Mg^{2+} ions together with the rest of the components needed for the LAMP reaction. However, we found that the buffer stock became turbid after 2–3 weeks of storage. Providing Mg^{2+} from a separate solution directly into the master mix circumvented the problem. Also it showed that providing Mg^{2+} from a $MgCl_2$ solution gave a slightly higher sensitivity as compared to $MgSO_4$. Another advantage of providing Mg^{2+} from a separate solution is that concentrations can be varied a little during optimization of sensitivity, rather than specificity.
4. The procedure for designing LAMP primers described here can be generalized to the design of LAMP primers. The net-based software used here can handle any DNA sequence <2,000 bp in FASTA format. If a text file of the sequence is used, the program will fill in gaps in order to produce a simulated FASTA format with the sequence split in columns of 10 nucleotide positions. This means that gaps will be introduced which the program counts as nucleotides. As a consequence, the highest number of nucleotides in a sequence will in fact be >1,800.
5. The Primer Explorer software provides a module for the design of loop primers. However, we were unable to get it working. In fact, the spacing between the ends of the F2/B2 primers and the F1c/B1c binding sites available for loop primer design are in most cases not more than 20–30 nucleotide positions. In most cases we have seen the number of possible primers in that space is restricted so that they can quite easily be designed manually.
6. The amplification reaction in a LAMP assay will generally run without loop primers. However, since loop primers bind to the loops which are formed as secondary structures in the amplified DNA, they can provide a higher number of the basic amplicons necessary to initiate the LAMP reaction. As a consequence of

this, the number of DNA molecules specifically synthesized per unit-time during LAMP is higher as compared to a reaction without loop primers. However, according to our experience not all loop primers have a positive effect on reaction time. This can be tested by comparison of the time needed to produce a fluorescent signal with a target DNA concentration in the range of the detection minimum by LAMP reactions with one of each loop primers and reactions with both loop primers and no loop primers.

7. We do not prepare the master mix on ice since we did not find any difference in the reaction performance. Stock solutions of the primers are kept at 4 °C and sufficient primer mix is produced for 80 LAMP reactions. Primer mixes are kept at 4 °C. Nucleotide mix is stored at –20 °C. Once an aliquot is thawed it is kept at 4 °C. All components stored at 4 °C are taken from the fridge just prior to preparation of the master mix. The *Bst* polymerase enzyme as well as the calcein amplification detection reagent are kept at –20 °C. Both components are diluted in glycerol to keep them fluid under the temperature conditions. We retrieve them from the freezer just before addition to the master mix and put them back at –20 °C immediately after use. Water is added from a supply stored at ambient temperature.
8. Isolation of highly purified genomic fungal DNA can be performed with mycelia grown in 100 ml of 3% (w/v) malt extract broth with 0.3% (w/v) soy peptone for 5 days at ambient temperature on a rotary shaker (120 rpm) in 250 ml Erlenmeyer flasks. Mycelia are vacuum filtered through paper filter discs, washed with 2 × 50 ml of sterile tap water and freeze dried. Lyophilized mycelia are finely ground with a mortar and pestle adding a small amount of sterile sea sand until no complete hyphae are seen under the microscope. Ground mycelium is subjected to DNA extraction as described by Lee and Taylor (25) with the following modifications: The ground mycelium from a 100 ml culture is subjected to 12 ml lysis buffer, containing 1% dithiotreitol instead of 2-mercaptoethanol. Extraction is done with phenol, phenol/CIA (chloroform:isoamylalcohol 24:1 v/v), and CIA, respectively, in three successive extractions by horizontally shaking the tubes for 15 min at ambient temperature. DNA is resuspended in ultrapure sterile water and aliquots stored at –20 °C. DNA working solutions are kept at 4 °C.
9. LAMP reactions can be run in any device capable of producing a constant temperature between 60 and 70 °C and the capacity to keep the temperature stable. We have run LAMP reactions in a thermal cycler, a heating block, a hybridization oven, a laboratory water bath and even in water of 65 °C kept in a

closed Styrofoam box, with comparable results. Others have even used a discardible pocket warmer as incubation device (see (26)).

10. Under daylight conditions or under a neon tube positive reactions appear in a greenish yellow tone, whereas negative reactions have a brownish rose color. Visualization of reactions against a light source shows that positive reactions become turbid in addition whereas negative reactions stay clear. Turbidity occurs from precipitation of a magnesium–pyrophosphate complex which is less water soluble as compared to the uncomplexed compounds.
11. For *F. graminearum* the most closely related species are *Fusarium* spp. within section Discolor (*F. bactridioides, F. buharicum, F. crookwellense, F. culmorum, F. flocciferum, F. heterosporum, F. lunulosporum, F. macroceras, F. reticulatum, F. robustum, F. sambucinum, F. sublunatum, F. sulphureum, F. trichothecioides, F. tumidum and F. venenatum*). We duely recommend to ask the help of a specialist in fungal taxonomy for information about the taxonomical position of the target fungus and to plan your experiments accordingly. Make sure to use only strains which come from an official culture collection specialized in the maintenance of fungi. If a ribosomal DNA sequence is available for the target fungus, use the BLAST algorithm available under http://www.blast.ncbi.nlm.nih.gov/ to obtain a pair wise alignment of homologues sequences. This may provide a rough overview over the taxonomical relationships between the target species and other fungi.
12. We found that the reaction runs at neglectable speed at ambient temperature. The reaction can therefore be stopped at any time during LAMP by cooling under cold tap water and restarted by restoring the optimal temperature conditions. Heating the reaction to 95 °C for 2 min will denature the *Bst* DNA polymerase enzyme and render it permanently inactive.
13. The length of incubation intervals can be deliberately varied according to the boundary conditions of the actual assay to be developed.
14. In general, fungal mycelia should be grown on agar medium and under temperature appropriate for optimal growth. Cultures should produce abundant aerial mycelia. According to our experience with *F. graminearum* and also with *F. tricinctum* it is not necessary for the cultures to sporulate. We also have had good results with cultures grown on suboptimal media provided isolates were able to grow.
15. Double volume of master mix is necessary to cover the grains completely and to allow for washing off all fungal material from the grains surface. Remove grains with a flamed forceps.

Make sure that forceps has cooled to ambient temperature to prevent heating of the master mix and premature start of the reaction or heat inactivation of the enzyme (see Note 11). Use a blank agar plate to cool forceps.

16. As shown in Fig. 5 there seem to be inhibitors of the LAMP reaction in some of the undiluted washings which, however, loose their influence at dilution. Sample washings were diluted between 10^{-1} and 10^{-6} because in some samples the concentration of inhibitors was high and fluorescence only occurred after massive dilution of the washing. No data for application of the F. graminearum specific LAMP in the analysis of bulk samples were available. Therefore, F. tricinctum specific primers were used to exemplify this kind of application. Prmers used are published in (27).

References

1. Mullis KB, Faloona FA (1987) Specific synthesis of DNA in vitro via a polymerase-catalyzed chain reaction. Methods Enzymol 155:335–350
2. Saiki RK, Gelfand DH, Stoffel S, Scharf SJ, Hguchi R, Horn GT, Mullis KB, Ehrlich HA (1988) Primer-directed enzymatic amplification of DNA with a thermostable DNA polymerase. Science 239:487–491
3. Niessen L (2008) PCR-based diagnosis and quantification of mycotoxin-producing fungi. Adv Food Nutr Res 54:81–137
4. Notomi T, Okayama H, Masubuchi H, Yonekawa T, Watanabe K, Amino N, Hase T (2000) Loop-mediated isothermal amplification of DNA. Nucleic Acids Res 28:e63
5. Nagamine K, Hase T, Notomi T (2002) Accelerated reaction by loop-mediated isothermal amplification using loop primers. Mol Cell Probes 16:223–229
6. Njiru ZK, Mikosza ASJ, Matovu E, Enyaru JCK, Ouma JO, Kibona SN, Thompson RCA, Ndung'u JM (2008) African trypanosomiasis: sensitive and rapid detection of the sub-genus Trypanozoon by loop-mediated isothermal amplification (LAMP) of parasite DNA. Int J Parasitol 38:589–599
7. Mori Y, Hirano T, Notomi T (2006) Sequence specific visual detection of LAMP reactions by addition of cationic polymers. BMC Biotechnol 6(3):10
8. Mori Y, Nagamine K, Tomita N, Notomi T (2001) Detection of loop-mediated isothermal amplification reaction by turbidity derived from magnesium pyrophosphate formation. Biochem Biophys Res Commun 289:150–154
9. Tomita N, Mori Y, Kanda H, Notomi T (2008) Loop-mediated isothermal amplification (LAMP) of gene sequences and simple visual detection of products. Nat Protoc 3:877–882
10. Mori Y, Kitao M, Tomita N, Notomi T (2004) Real-time turbidimetry of LAMP reaction for quantifying template DNA. J Biochem Biophys Methods 59:145–157
11. Endo S, Komori T, Ricci G, Sano A, Yokoyama K, Ohori A, Kamai K, Franco M, Miyaji M, Nishimura K (2004) Detection of gp43 of *Paracoccidioides brasiliensis* by the loop-mediated isothermal amplification (LAMP) method. FEMS Microbiol Lett 234:93–97
12. Ohori A, Endo Sano A, Yokoyama K, Yarita K, Yamaguchi M, Kamai K, Miyaji M, Nishimura K (2006) Rapid identification of *Ochroconis gallopava* by a loop-mediated isothermal amplification (LAMP) method. Vet Microbiol 114:359–365
13. Tomlinson JA, Barker I, Boonham N (2007) Faster, simpler, more-specific methods for improved molecular detection of *Phytophthora ramorum* in the field. Appl Environ Microbiol 73:4040–4047
14. Hayashi N, Arai R, Tada S, Taguchi H, Ogawa Y (2007) Detection and identification of *Brettanomyces/Dekkera* sp. yeasts with a loop-mediated isothermal amplification method. Food Microbiol 24:778–785
15. Gadkar V, Rillig MC (2008) Evaluation of loop-mediated isothermal amplification

(LAMP) to rapidly detect arbuscular mycorrhizal fungi. Soil Biol Biochem 40:540–543

16. Sun J, Najafzadeh MJ, Vicente V, Xi L, de Hoog GS (2010) Rapid detection of pathogenic fungi using loop-mediated isothermal amplification, exemplified by Fonsecaea agents of chromoblastomycosis. J Microbiol Meth 80:19–24
17. Sun J, Li X, Zeng H, Zhi Z, Lu C, Xi L, de Hoog GS (2010) Development and evaluation of loop-mediated isothermal amplification(LAMP) for the rapid diagnosis of *Penicillium marneffei* in archived tissue samples. FEMS Immunol Med Microbiol 58:381–388
18. Lucas S, da Luz Martins M, Flores O, Meyer W, Spencer-Martins I, Inácio J (2010) Differentiation of *Cryptococcus neoformans* varieties and *Cryptococcus gattii* using CAP59-based loop-mediated isothermal DNA amplification. Clin Microbiol Infect 16:711–714
19. Matsuzawa T, Tanaka R, Horie Y, Gonoi T, Yaguchi T (2010) Development of rapid and specific molecular discrimination methods fro pathogenic *Emericella* species. Nihon Ishinkin Gakkai Zasshi 51:109–116
20. Tomlinson JA, Dickinson MJ, Boonham N (2010) Detection of *Botrytis cinerea* by loop-mediated isothermal amplification. Lett Appl Microbiol 51:650–657
21. Niessen L, Vogel RF (2010) Detection of Fusarium graminearum DNA using a loop-mediated isothermal amplification (LAMP) assay. Int J Food Microbiol 140:183–191
22. Doohan FM, Brennan J, Cook BM (2003) Influence of climatic factors on *Fusarium* species pathogenic to cereals. Eur J Plant Pathol 109:755–768
23. Hippeli S, Elstner EF (2002) Are hydrophobins and/or non specific lipid transfer proteins (ns-LTPs) responsible for gushing in beer? New hypotheses on the chemical nature of gushing inducing factors. Z Naturforsch C 57:1–8
24. Nirenberg H (1976) Untersuchungen über die morphologische und biologische Differenzierung in der *Fusarium*-Sektion Liseola. Mitt Biolog Bundesanstalt für Land-u Forstw (Berlin-Dahlem) 169:117
25. Lee SB, Taylor JW (1990) Isolation of DNA from fungal mycelia and single spores. In: Innis MA, Gelfand DH, Sninsky JJ, White TJ (eds) PCR protocols: a guide to methods and applications. Academic Press, New York, pp 282–287
26. Hatano B, Maki T, Obara T, Fukumoto H, Hagisawa K, Matsushita Y, Okutani A, Bazartseren B, Inoue S, Sata T, Katano H (2010) LAMP using a disposable pocket warmer for anthrax detection, a highly mobile and reliable method for anti-bioterrorism. Jpn J Infect Dis 63:36–40
27. Niessen L, Gräfenhan T, Vogel RF (2012) ATP citrate lyase 1 (acl1) gene-based loop-mediated isothermal amplification assay for the detection of the Fusarium tricinctum species complex in pure cultures and in cereal samples. Int J Food Microbiol 158:171–185

Chapter 15

Multiplex-Tandem PCR for Fungal Diagnostics

Anna Lau, Keith Stanley, and Tania Sorrell

Abstract

Multiplex, real-time PCR has become an invaluable tool for the rapid identification of pathogens in clinical specimens enabling earlier and more targeted management of antimicrobial therapy. In this chapter, we describe the methodology behind a novel multiplex-tandem PCR (MT-PCR) platform designed for the rapid identification of up to 18 fungal pathogens in blood cultures, primary isolation plates, and whole blood, serum, and plasma.

Key words: MT-PCR, Multiplex PCR, Fungi, Candida, Blood

1. Introduction

Molecular platforms are an integral part of today's clinical microbiology laboratory providing expedited species identification and improved diagnostic sensitivity and accuracy. However, the in-house development of most assays and the lack of standardized methods have prevented their implementation into routine clinical practice and the conduct of large-scale inter-laboratory validations. The recent introduction of commercially available systems such as the LightCycler SeptiFast assay (Roche Diagnostics, Mannheim, Germany), may promote standardization of molecular methods for diagnosing invasive fungal infections. In this chapter, the methodology behind a novel multiplex-tandem PCR (MT-PCR) platform designed for the rapid identification of fungal pathogens in blood cultures (1), primary isolation plates (2), and whole blood, serum, and plasma (3) is described. These assays were developed as collaboration between the Centre for Infectious Diseases and

Louise O'Connor and Barry Glynn (eds.), *Fungal Diagnostics: Methods and Protocols*, Methods in Molecular Biology, vol. 968, DOI 10.1007/978-1-62703-257-5_15, © Springer Science+Business Media New York 2013

Microbiology, University of Sydney, Westmead, New South Wales and AusDiagnostics.

MT-PCR (trade mark *Easy*-Plex™) is a novel, two-step, nested, real-time PCR that can simultaneously identify up to 72 different targets in a single assay (4). After nucleic acid extraction, DNA is amplified in a multiplex PCR reaction. The number of amplification cycles is kept short (10 cycles) to enable the enrichment of amplicons without creating competition between different primer sets. Following this first round of amplification, the product is diluted 1:100 and aliquoted into wells of the 72-well *Easy*-Plex ring. Each well contains lyophilized, target-specific primers that have been designed to nest within those used in the first amplification. In this way, high specificity is maintained because target DNA must bind to both the first and second round set of primers in order to be detected. Fluorescence is measured at the end of each extension step using Eva-Green dye that binds to double-stranded DNA. After cycling, a melt-curve analysis is performed. In-built analysis software compares the resulting melt temperature of the amplified product with the expected melt temperature of the target, to determine the presence or absence of an organism. An internal PCR positive control, known as the "spike," is included with each sample to monitor PCR efficiency and inhibition.

MT-PCR kits are distributed by AusDiagnostics. Each kit contains mastermix, tubes containing lyophilized primers for the first round (multiplex) PCR and 72-well *Easy*-plex rings containing lyophilized primers for the second round (target specific) PCR. All reagent tubes are labeled and color-coded to enable easy set-up. For high-throughput laboratories, a robot is available to automatically add sample, perform multiplex PCR and aliquot amplicon-mastermix solution into the gene disc, thus providing a walk-away system that is easy to operate.

In addition to identifying fungal pathogens, several other kits have been developed for routine use in clinical microbiology laboratories. Details are located on the AusDiagnostics website (http://www.ausdiagnostics.com).

2. Materials

Prior to any experimental procedure, all work surfaces and equipment including cabinets, pipettes, and racks should be cleaned with NucleoClean decontamination solution (Chemicon International, Temecula, CA) (see Note 1). All procedures involving DNA manipulation should be performed in a Class II Biological Safety Cabinet. The three steps of DNA extraction, PCR set-up and aliquoting of diluted amplicons into the *Easy*-Plex ring should be performed in separate, independently equipped laboratories that are set out in a unidirectional workflow to prevent carry over contamination. Gloves and sterile plugged pipette tips should be used for all molecular procedures. Reagents should be prepared using ultrapure water (prepared by purifying deionized water to a sensitivity of 18 MΩ cm at 25°C) and analytical grade reagents. Reagents and solutions should be stored at room temperature unless otherwise indicated.

2.1. Nucleic Acid Extraction

1. Alkali wash solution: 0.5 M NaOH and 0.5 M tri-sodium citrate dehydrate. Dissolve 20 g NaOH and 12.9 g tri-sodium citrate dehydrate in 1 L ultrapure water. Divide solution into five 200 ml bottles. Autoclave solution at 121°C for 15 min. Cool before use.
2. Phosphate-buffered saline: Dissolve 8 g NaCl, 1.21 g K_2HPO_4, and 0.34 g KH_2PO_4 in 1 L ultrapure water. Autoclave solution at 121°C for 15 min. Cool before use.
3. Lyticase: Dissolve powder in molecular biology grade water to give a final concentration of 25 U/μl. Aliquot 100 μl of solution into sterile eppendorf tubes. Store solution at −20°C until needed (see Note 2).
4. NucliSENS easyMAG instrument and reagents (bioMerieux, Marcy l'Etoile, France).
5. EDTA solution (0.5 M; pH 7.4): Dissolve 146.12 g EDTA·$2H_2O$ in 800 ml ultrapure water. Adjust pH to 7.4 using NaOH whilst stirring. Make the volume up to 1 L. Autoclave at 121°C for 15 min. Store at 4°C.
6. Erythrocyte lysis buffer: 0.155 M NH_4Cl, 0.01 M NH_4HCO_3, 0.1 mM EDTA [pH 7.4]. Dissolve 8.829 g NH_4Cl, 0.7906 g NH_4HCO_3, and 200 μl EDTA (0.5 M) in 1 L ultrapure water. Autoclave at 121°C for 15 min.
7. High Pure PCR template preparation kit (Roche Diagnostics).

2.2. MT-PCR

All MT-PCR (or *Easy*-Plex) reagents, tubes, and *Easy*-Plex rings containing lyophilized primers are provided in each kit (AusDiagnostics). Reagents should be stored at −20°C until required. Tubes and *Easy*-plex rings can be stored at room temperature or at 4°C depending on laboratory space. The first multiplexed amplification reaction can be performed on any thermocycler. Due to the nature of the ring-based system, the second amplification needs to be performed on the Rotorgene 6000 (Qiagen). Computer analysis software is provided by AusDiagnostics. A robot is an available option for high-throughput laboratories.

3. Methods

3.1. Nucleic Acid Extraction from Blood Cultures

1. Mix the fungal positive blood culture well and wipe the top of the bottle with an alcohol swab (see Note 3). Using a sterile needle and syringe, carefully transfer 100 μl of blood culture to a 2 ml sterile microcentrifuge tube. Do not recap the syringe. Dispose of it in a sharps container.
2. Add 1 ml of alkali wash solution to the tube containing 100 μl blood culture (see Note 4). Vortex at high speed for 15 s and incubate at room temperature for 5 min.
3. Centrifuge the mixture at 16,000 × *g* for 5 min. Gently remove and dispose of the supernatant using a fine tip sterile Pasteur pipette being careful not to disturb the cell pellet at the bottom of the tube (see Note 5).
4. Resuspend the pellet in 1 ml PBS solution. Vortex thoroughly for 15 s and centrifuge at 16,000 × *g* for 5 min. Carefully remove and dispose of the supernatant using a fine tip sterile Pasteur pipette being careful not to disturb the cell pellet (see Note 5).
5. Resuspend the pellet in 200 μl molecular biology grade water and add 2 μl Lyticase (50 U). Vortex thoroughly for 15 s and incubate the suspension at 37°C for 1 h.
6. Pulse-spin the tube for 5 s to remove condensation. Using a sterile pipette, transfer the entire suspension to a nucliSENS easyMAG lysis tube. Vortex for 10 s and incubate at room temperature for 10 min.
7. Prepare the magnetic bead solution by adding molecular biology grade water to the nucliSENS easyMAG Magnetic Silica Beads using the automated pipette at setting 1 supplied by bioMerieux. Dispense aliquots of the magnetic bead solution into plastic wells using the automated pipette at setting 2.
8. Transfer the lysis solution containing the sample to the disposable tray provided in the nucliSENS easyMAG kit. Add the aliquots of the magnetic bead solution using the automated pipette at setting 3.
9. On the nucliSENS easyMAG instrument, select 250 μl as input volume, whole blood as sample type and 60 μl as elution volume. The nucliSENS easyMAG instrument will automatically isolate nucleic acid. After the run has finished, transfer the eluted DNA to a sterile 2 ml microcentrifuge tube (see Note 6) and store at −20°C until required.

3.2. Nucleic Acid Extraction from Whole Blood

Nucleic acid is extracted from whole blood using the High Pure PCR Template Preparation Kit with some minor modifications listed below (see Note 7).

1. Separate the specimen into 3 × 1 ml samples in 2 ml microcentrifuge tubes.
2. Add 1 ml erythrocyte lysis buffer, mix by inversion, and freeze at −20°C for 10 min. Centrifuge at 4,000 × *g* for 10 min. Gently remove the supernatant using a fine tip Pasteur pipette. Repeat this step to ensure removal of all erythrocytes (see Note 8).
3. Spheroplasts were combined and resuspended in 1 ml molecular biology grade water, vortexed for 15 s and centrifuged at 5,400 × *g* for 5 min.
4. The supernatant was removed and the pellet was incubated in 300 μl tissue lysis buffer and 60 μl proteinase K at 55°C for 1 h to overnight.
5. DNA was purified using spin-columns. In our experience, the best results were obtained using the High Pure PCR Template Preparation Kit (Roche Diagnostics).
6. DNA was eluted in a 60 μl volume.

3.3. Nucleic Acid Extraction from Serum and Plasma

1. To collect serum, centrifuge a plain blood collection tube (red top) containing clotted blood at 1,258 × *g* for 10 min. Gently transfer 1 ml of the serum to nucliSENS easyMAG lysis tube being careful not to disturb the gel layer at the bottom of the collection tube.
2. To collect plasma, repeat step 1 using a blood sample collected in an EDTA-coated blood collection tube (purple top).
3. On the nucliSENS easyMAG instrument, select either serum or plasma as the specimen type, 1 ml as the sample volume and 60 μl as elution volume. After the run has finished, transfer the eluted DNA to a sterile 2 ml eppendorf tube (see Note 6) and store at −20°C until required.

3.4. Preparation of Fungal Cultures for MT-PCR

Neither a pure culture nor nucleic acid extraction is required for the identification of yeasts or molds from primary isolation agar plates using MT-PCR. Simply take a sweep of growth and place into a 2 ml microcentrifuge tube containing 500 μl sterile water. Vortex for 15 s. The suspension is now ready for MT-PCR.

3.5. MT-PCR Amplification and Detection

Defrost first and second round mastermix solutions (provided in the MT-PCR kit) for each respective step. Vortex for 10 s and pulse spin before use (see Note 9). Always set up PCR reactions on ice.

1. First-round multiplex pre-amplification: Perform PCRs in a 20 μl volume consisting of 10 μl step 1 mastermix (AusDiagnostics), 8 μl molecular biology grade water, and 2 μl sample DNA (see Note 10). Prepare reactions in the 200 μl PCR tubes containing lyophilized primers supplied in the MT-PCR kit. Transfer the tubes to the 36-well rotor and place in the Rotogene 6000 thermal cycler. The MT-PCR step 1 is pre-programmed to cycle under the following conditions: 95°C for 5 min, followed by 15 cycles of 95°C for 10 s, 60°C for 20 s, and 72°C for 20 s.
2. Second-round quantification amplification: Prepare mastermix for each sample in a clean laboratory by combining 150 μl step 2 mastermix (AusDiagnostics) and 146 μl molecular biology grade water. Transfer this solution to the post-PCR laboratory in preparation to adding to wells of the *Easy*-Plex ring. Pulse-spin the PCR products from step 1 to remove residual condensation. Add 4 μl of each PCR product to its respective mastermix tube and mix by pipetting up and down. Aliquot 20 μl into the wells of the 72-well *Easy*-Plex ring containing species-specific lyophilized primers that are internalized to those used in the multiplex PCR. Seal the disc using the gene disc sealer (AusDiagnostics), insert into the 72-well disc rotor and place in the Rotorgene 6000 thermal cycler. The MT-PCR step 2 is pre-programmed to cycle at 95°C for 10 min, followed by 30 cycles of 95°C 30 s, 60°C for 10 s, and 72°C for 10 s. Fluorescence is automatically measured at the end of each 72°C extension step. Following cycling, a melting curve is automatically generated and in-built analysis software compares the given melt temperature with the expected melt temperature to determine the presence or absence of an organism. Easy to interpret, tabular reports are automatically generated for each sample.
3. For laboratories equipped with an *Easy*-Plex robot, the robot automatically performs the entire step 1 process and the addition of samples into the 72-well *Easy*-Plex ring. The user needs only to seal the *Easy*-Plex ring, insert it into the rotor and place in the Rotorgene 6000 thermal cycler. AusDiagnostics also sell a positive control consisting of synthetic targets for each of the fungal ID targets.

4. Notes

1. NucleoClean decontamination solution can be substituted for 1% bleach. Wipe with 70% ethanol to prevent bleach de-colorization.
2. Lyticase will not fully dissolve in solution and a pellet is often present. Ensure the solution is thoroughly vortexed before adding into samples.

3. Ensure that the top of the blood culture bottle is properly cleansed with the alcohol swab to remove residual blood and prevent test contamination.
4. The blood culture will turn green after the addition of alkali wash solution.
5. If the cell pellet is disturbed, expel the contents of the pipette back into the eppendorf tube and repeat centrifugation.
6. Be careful not to disturb the magnetic pellet on the side of the well when transferring the eluate.
7. To maximize the detection of the fungal DNA, erythrocytes are removed and white blood cells are concentrated.
8. If the cell pellet is still red after the second wash, repeat for a third time. However, note that there may be further loss of material if this additional step is performed impacting on the sensitivity of the assay.
9. Be careful not to over centrifuge the mastermix as the dye can collect at the bottom of the tube.
10. For the amplification of fungal DNA from whole blood, the **step 1** mastermix volume should be reduced to 8 μl and the input DNA volume increased to 11 μl.

References

1. Lau A et al (2008) Multiplex tandem PCR: a novel platform for rapid detection and identification of fungal pathogens from blood culture specimens. J Clin Microbiol 46(9):3021–3027
2. u A et al (2008) Colony multiplex-tandem PCR for rapid, accurate identification of fungal cultures. J Clin Microbiol 46(12):4058–4060
3. Lau A et al (2010) Comparison of whole blood, serum and plasma for early detection of candidemia by multiplex-tandem PCR. J Clin Microbiol 48(3):811–816
4. Stanley KK, Szewczuk E (2005) Multiplexed tandem PCR: gene profiling from small amounts of RNA using SYBR Green detection. Nucleic Acids Res 33(20):e180

Chapter 16

Specific Detection of *Pneumocystis jirovecii* in Clinical Samples by Flow Cytometry

Joana Barbosa, Sofia Costa-de-Oliveira, Ana Teresa Silva, Acácio Gonçalves Rodrigues, and Cidália Pina-Vaz

Abstract

Pneumocystis jirovecii is an opportunistic pathogen responsible for severe pneumonia in immunocompromised patients. Flow cytometry (FC) is a method widely used in different areas of clinical diagnosis like hematology and immunology. Recently it has started to be used in microbiology with a great potential for diagnosis of emergent microorganisms in clinical samples, especially when present in low numbers. The detection of *Pneumocystis jirovecii* in respiratory samples can be performed by FC, using specific monoclonal antibodies. Considering clinical diagnosis as a reference method, we previously showed FC to be 100% sensitive and specific when compared to immunofluorescence. Being an automated method, it is faster and less subject to human error than microscopic evaluation.

Key words: Flow cytometry, Respiratory samples, *Pneumocystis jirovecii* diagnosis, Specific monoclonal antibodies, Emergent fungi, Pneumonia, Immunocompromised patients

1. Introduction

Pneumocystis jirovecii is an atypical microorganism classified until 2001 as a protozoan and, since then, as fungi (1). Ribosomal DNA and RNA studies revealed similarities with fungi, although its diagnostic form is a cyst and is impossible to cultivate in vitro. In humans, this pathogen is responsible for an opportunistic infection in the lower respiratory tract leading to a severe pneumonia in immunocompromised patients (2, 3). Identification of the organism has for a long time been based upon direct microscopy either using Gomori Grocott or immunofluorescent staining. Both procedures are time consuming, show low sensitivity, and are subject

Louise O'Connor and Barry Glynn (eds.), *Fungal Diagnostics: Methods and Protocols*, Methods in Molecular Biology, vol. 968, DOI 10.1007/978-1-62703-257-5_16, © Springer Science+Business Media New York 2013

to human error especially when samples yield a low number of organisms (4–7). Molecular biology techniques, particularly polymerase chain reaction, result in a higher sensitivity and specificity when compared to the microscopic detection in respiratory samples (8, 9) and provide a faster result by diagnostic microbiological laboratories. However, this molecular approach is not available at many clinical microbiology laboratories (10, 11). Flow cytometry was developed during the 1960s and, ever since, it has had a considerable impact in hematology and immunology since this method allows both quantification of individual cells and morpho-functional evaluation. Whenever cells suspended in a liquid ("sheat fluid") are exposed individually through a laser beam, the light scattered and the fluorescence emitted by illuminated cells are measured and those results will represent cumulative cytometric characteristics. The scattered light provides intrinsic cell information like size and complexity (12–15). When using fluorochromes, a wide range of cellular physiological or morphological parameters can be additionally evaluated. Also, whenever combined with specific antibodies, it can be used to detect specific microorganisms present in low concentration in clinical specimens.

This method has contributed to a more detailed insight into cellular phenomena, which are very difficult to study with other alternative methods (15–17). FC offers a broad and growing range of potential applications in clinical microbiology (13). When compared with conventional microscopy, this method is faster, more objective, and automated (18). Presently, distinct applications of flow cytometry in clinical microbiology have been developed by our research group, showing considerable advantages over classic methods mainly in mycology (19–24), bacteriology (25, 26), and parasitology (27, 28). We now describe a flow cytometric protocol for specific detection of *Pneumocystis jirovecii* in clinical respiratory samples, with high sensitivity and specificity.

2. Materials

1. *Respiratory samples*: Bronchoalveloar lavage fluid (BAL 30 ml), bronchial washing fluid (BW 30 ml), bronchial secretions (BS 2–4 ml).
2. Mucolytic agent *N*-acetyl-L-cysteine (Merck).
3. Filter (30-μm pore size–Partec CellTrics®).
4. Sterile distilled water.
5. Positive and negative respiratory sample should be used as control for FC analysis, since there are no commercialized controls available.

6. Detect IF™ kit *Pneumocystis carinii* (Axis-Shield Diagnostics Limited, UK) with *P.carinii* mouse monoclonal antibody (1 ml), FITC-conjugated antibody (2 ml), enzyme diluent (3 ml).
7. Microscope glass slides and cover slips (22 × 40 × 0.15) (Fisher, Pittsburgh, PA).
8. Epifluorescence microscope Leitz Laborlux K (Leica, NY, USA).
9. FACSCalibur Flow Cytometer (BD Biosciences, Sydney, Australia), standard model equipped with three photomultipliers (PMTs), standard filters (FL1: BP 530/30 nm; FL2: BP 585/42 nm; FL3: LP 650 nm), a 15-mW 488-nm argon laser and with CellQuest Pro Software (version 4.0.2, BD Biosciences, Sydney, Australia).
10. BDFacsFlow, BD Facs Rinse, and BDFacs Clean solutions (BD Biosciences, Sydney, Australia).
11. Rainbow calibration particles (6 peaks), 3.0–3.4 μm, for calibration (BD Biosciences, Sydney, Australia).

3. Methods

Carry out all procedures at room temperature, unless otherwise specified.

3.1. Respiratory Sample Preparation

1. For all clinical samples, mix mucolytic agent *N*-acetyl-L-cysteine with equal volume of sample (for BAL and BW, 30 ml; for BS, 2–4 ml) (see Note 1).
2. Incubate in a water bath at 37°C for 20 min or until completely dissolved, followed by centrifugation at 3,000 × *g* for 15 min.
3. Wash twice with 1 ml of distilled sterile water, centrifuge as described in step 2 above.
4. After the last centrifugation, aspirate the supernatant until 100 μl of sediment remains. Use this suspension for FC analysis.

3.2. Sample Staining for Flow Cytometry

To obtain the cell suspension:

1. Filter the sediment through a 30-μm pore size filter in order to remove large particles (see Note 2) to avoid technical problems when aspirating specimens through the FC aspiration needle.

2. Mix 20 μl of specific enzyme (Detect IF™ kit *P. carinii*) with 100 μl of the respiratory sample, and incubate at 37°C for 30 min (see Note 3).
3. Following incubation, add 500 μl of sterile H_2O and centrifuge at 3,000 × *g* for 5 min.
4. Remove the supernatant and mix 100 μl of the sediment with 10 μl of specific *P. jirovecii* mouse monoclonal antibody. Incubate at 37°C for 15 min in the dark (see Note 4).
5. Increase the mixture volume by adding 500 μl of sterile H_2O and centrifuge at 3,000 × *g* for 5 min.
6. Remove the supernatant and mix 100 μl of the sediment with 10 μl of FITC-conjugated antibody. Incubate at 37°C for 15 min in the dark.
7. Remove the supernatant and resuspend the sediment in 100 μl of sterile H_2O.
8. Vortex for 30 s and transfer to a propylene tube for analysis in the Flow Cytometer.

3.3. Flow Cytometry

For the reader an introduction to flow cytometry-specific terms is given (see Note 5).

3.3.1. Flow Cytometry Acquisition and Analysis

The optical characteristics of the cell suspensions are evaluated on a FACSCalibur Flow Cytometer standard model equipped with three photomultipliers (PMTs), standard filters (FL1: BP 530/30 nm; FL2: BP 585/42 nm; FL3: LP 650 nm), a 15-mW 488-nm argon laser and with CellQuest Pro Software (version 4.0.2, BD Biosciences, Sydney, Australia). Operating conditions included log scales on all detectors (forward scatter [FSC], side scatter [SSC], and fluorescence detectors [FL1]).

1. Turn on the equipment (FC), computer and open the CellQuest Pro Software for instruments control settings and definition of plots for data collection.
2. Open the instruments controls and use 488 nm as the excitation wavelength (laser) to excite the fluorescence from FITC (or many other possible dyes).
3. Select a histogram on the plots list and use forward scatter [FSC] and/or side scatter [SSC] signals for triggering scattergrams. A software gate excluding low-level scatter signals may be set to remove events due to noise and particulate contaminants in samples (see Note 6).
4. Adjust the PMTs voltage to the first logarithmic (log) decade in the histogram (Fig. 1a) for definition of the acquisition settings:

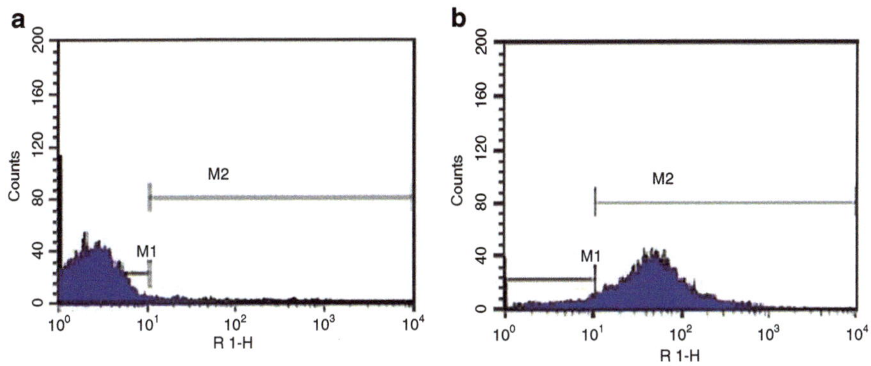

Fig. 1. Histogram correlating counts *versus* fluorescence intensity displayed by *Pneumocystis jirovecii*, and registered at FL1 (green fluorescence, 525 nm), from **(a)** a non-stained respiratory sample, representing autofluorescence; **(b)** a positive respiratory sample stained with 10.0 μg/ml of specific monoclonal antibody. Regions M1 and M2 correspond to fluorescence intensity of non-stained samples and to specific antibody fluorescence, respectively.

(a) For *P. jirovecii* analysis, instruments settings should be as follows: FSC (forward scatter) E01; SSC (side scatter) 600 V; fluorescence detector FL1 690 V; SSC threshold 52 V.

5. Select a histogram correlating the detected fluorescence FL1, on the *X*-axis, and counts of acquired events, on the *Y*-axis.
6. Select a dot plot correlating the detected fluorescence FL1, on the *X*-axis, and side scatter [SSC] signals, on the *Y*-axis.
7. With a non-stained sample (see Note 7) (representing autofluorescence in the sample), insert the suspension into the aspiration needle and select the low rate acquisition velocity.
8. During the acquisition, on the scattergram define the specific region for analysis, where during the acquisition (at least 50,000 cells) (see Note 8), captured cells are concentrated.
9. On the histogram, define a gate (fluorescence signal) at the end of autofluorescence zone of each sample (M1) and another (M2) for the evens showing fluorescence level higher than the limit of such gate at FL1 (green fluorescence) (see Note 9).
10. Check the acquisition zone of stained cells showing high fluorescence. For *P. jirovecii*-specific monoclonal antibody, as the mean value of fluorescence at FL1 increases, the signal intensity moves to the second logarithmic scale, when compared with the corresponding value for the autofluorescence (Fig.1b).
11. On the dot plot, define the specific region where stained cells are acquired. Samples showing events on the autofluorescence acquisition gate should be considered negative for *P. jirovecii* (Fig. 2a); samples showing events on the right to autofluorescence acquisition gate should be considered positive (Fig. 2b).

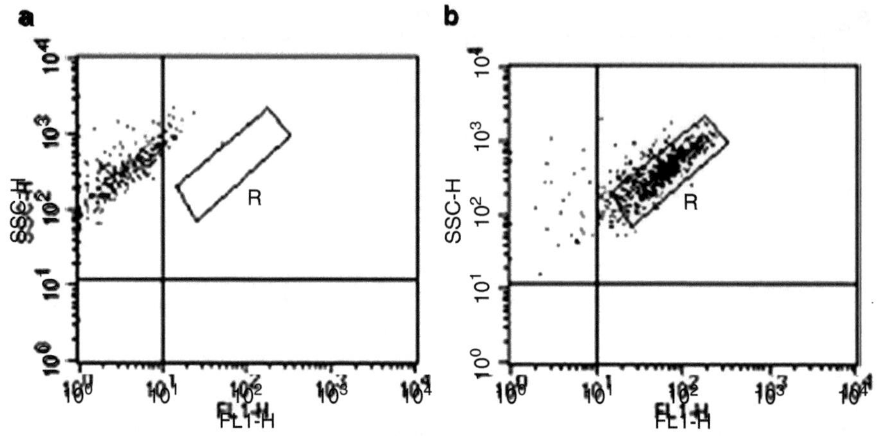

Fig. 2. Two dimensional dot-plot correlating side scatter (SSC) *versus* FL1 (*green* fluorescence, 525 nm), fluorescence intensity displayed by a negative **(a)** and a positive **(b)** respiratory samples stained with 10.0 μg/ml of specific *Pneumocystis jirovecii* monoclonal antibody. Regions R correspond to acquisition gate of stained cysts with specific antibody.

4. Notes

1. For Bronchoalveloar lavage fluid (BAL) and bronchial washing fluid (BW) 30 ml should be used. For bronchial secretions (BS) 2–4 ml should be used. *N*-acetyl-L-cysteine is a mucolytic agent used for the liquefaction and decontamination of such bronchial samples.
2. For an FC analysis, cells or particles (the general term for any of the objects flowing through the Flow Cytometer) should be suspended in a buffer solution, and allowed to flow through the cytometer one by one.
3. At this step the specific enzyme used digests cellular tissue and will release *P. jirovecii* cells present in the sample allowing the subsequent antibody/antigen reaction to proceed. It is important to note that this indirect immunofluorescence uses two antibodies: the unlabeled first (primary) antibody specifically binds the target molecule and the secondary antibody, which carries the fluorophore, recognizes the primary antibody and binds to it.
4. It is not so critical to incubate in the dark but normally in our laboratory we protect from light in order to avoid any degradation of the sample.
5. Specific flow cytometry terms:
   (a) Autofluorescence: natural background fluorescence from particles and detected by the cytometer, usually showing low intensity of fluorescence.

(b) Fluorescence dyes: chemical compounds that absorbs light of a certain wavelength and then emits in a different wavelength, usually of a longer wavelength, being selected by a filter. Example: FITC 488 nm (excitation wavelength), 525 nm (emitted wavelength), green (emitted color), FL1 (5 nm).

(c) Laser and illumination cells, in a stream of fluid (suspension) move through the light beam, with only one particle at a time being illuminated with the light from a laser. The resulting illumination is bright enough to scatter the light and/or excite the fluorochrome making the cells fluorescent.

(d) Event: a term used to indicate anything that has been interpreted by the cytometer, correctly or incorrectly, as a single particle.

(e) Data analysis: the light emitted from each particle will be detected and stored in a data file for subsequent analysis. The data is normally displayed as histograms or dot plots.

  (i) Histogram: used to display the distribution of one parameter over the cells in the data file; the *X*-axis shows the intensity of the detected signal and the *Y*-axis measures the number of events counted. This can be also used to display data from FC experiments.

  (ii) Dot plot: two dimensional plots used to display of correlated data, by using two parameters to graph the data generated by flow analysis, with each dot representing the passage of one cell through the detector. The X- and Y axes measure different emissions, displaying a dot for each cell with a particular emission. A cell of a particular population type will show up as a dot in the quadrant of the dot plot designated for that population.

6. During the acquisition, the cells from the sample tube are injected into the sheath stream and flowed through a laminar flow cell. When light from a laser "interrogates" a cell that is passing through, light is scattered in all directions and this light can travel from the interrogation point down a path to a detector. The light that is scattered in the *forward* direction (along the same axis the laser is traveling) is detected in the Forward Scatter Channel (FSC) and the intensity of the signal will be attributed to cell size and cell membrane shape. The light that is scattered at 90° to the axis of the laser path is detected in the side scatter channel (SSC) and the intensity of this signal is proportional to the amount of cytosolic structure in the cell (Complexity). Since FSC is comparable to size and cell membrane shape and SSC to internal cell structure, a correlated

measurement between them can allow for differentiation of cell types in a heterogeneous cell population.

7. The non-stained sample is the filtered sample which has not undergone staining. Execute all the steps in Subheading 3.1 and Subheading 3.2 step 1 but do not perform the staining. This non-stained sample is important to guarantee the elimination of the natural background fluorescence from cells and detection of signal of specific stained sample.
8. We use 50,000 in order to guarantee a reasonable cell number (10E4) for sample evaluation.
9. When we define a gate we simulate isolation of a subset of cells on a plot allowing the ability to look at parameters specific to only that subset. For example if we have stained cells in the sample (M2), the cells will have a mean fluorescence intensity higher than the non-stained sample (autofluorescence) (M1).

References

1. Murray PR, Rosenthal KS, Pfaller MA (2005) Fungal classification, structure, and replication. In: Murray PR, Rosenthal KS, Pfaller MA (eds) Medical microbiology, 5th edn. Mosby, Philadelphia, Pennsylvania, pp 67–73
2. Su YS, Lu JJ, Perng CL, Chang FY (2008) *Pneumocystis jirovecii* pneumonia in patients with and without human immunodeficiency virus infection. J Microbiol Immunol Infect 41:478–4823
3. Kelley CF, Checkley W, Mannino DM, Franco-Paredes C, Del Rio C, Holguin F (2009) Trends in hospitalizations for AIDS associated *Pneumocystis jirovecii* pneumonia in the United States (1986 to 2005). Chest 136:190–197
4. Procop GW, Haddad S, Quinn J, Wilson ML, Henshaw NG, Reller LB, Artymyshyn RL, Katanik MT, Weinstein MP (2004) Detection of *Pneumocystis jirovecii* in respiratory specimens by four staining methods. J Clin Microbiol 42:3333–3335
5. Nevez G, Chabé M, Rabodonirina M, Virmaux M, Dei-Cas E, Hauser PM, Totet A (2008) Nosocomial *Pneumocystis jirovecii* infections. Parasite 15:359–365
6. Morris A, Netravali M, Kling HM, Shipley T, Ross T, Sciurba FC, Norris KA (2008) Relationship of *Pneumocystis* antibody response to severity of chronic obstructive pulmonary disease. Clin Infect Dis 47:e64–e68
7. Krajicek BJ, Thomas CF Jr, Limper AH (2009) *Pneumocystis* pneumonia: current concepts in pathogenesis, diagnosis, and treatment. Clin Chest Med 30:265–278
8. Durand-Joly I, Chabé M, Soula F, Delhaes L, Camus D, Dei-Cas E (2005) Molecular diagnosis of *Pneumocystis* pneumonia. FEMS Immunol Med Microbiol 45:405–410
9. Jiancheng W, Minjun H, Yi-jun A, Lan S, Zengzhu G, Jianrong S, Xixiong K (2009) Screening *Pneumocystis carinii* pneumonia in non-HIV-infected immunocompromised patients using polymerase chain reaction. Diagn Microbiol Infect Dis 64:396–401
10. Alvarez-Martínez MJ, Miró JM, Valls ME, Moreno A, Rivas PV, Solé M, Benito N, Domingo P, Muñoz C, Rivera E, Zar HJ, Wissmann G, Diehl AR, Prolla JC, de Anta MT, Gatell JM, Wilson PE, Meshnick SR (2006) Sensitivity and specificity of nested and real-time PCR for the detection of *Pneumocystis jirovecii* in clinical specimens. Diagn Microbiol Infect Dis 56:153–160
11. Robberts FJL, Liebowitz LD, Chalkley LJ (2007) Polymerase chain reaction detection of *Pneumocystis jirovecii*: evaluation of 9 assays. Diagn Microbiol Infect Dis 58:385–392
12. Watson JV (1991) Introduction to flow cytometry, 1st edn. Cambridge University Press, Cambridge
13. Álvarez-Barrientos A, Arroyo J, Cantón R, Nombela C, Sánchez-Pérez M (2000) Applications of flow cytometry to clinical microbiology. Clin Microbiol Rev 13:167–195
14. Givan A (2005) Flow cytometry: an introduction. In: Hawley TS, Hawley RG (eds) Methods in molecular biology: flow cytometry protocols, 2nd edn. Humana Press, Totowa, NJ, pp 1–32

15. Shapiro HM, Nebe-von-caron G (2005) Multiparameter flow cytometry of bacteria. In: Hawley TS, Hawley RG (eds) Methods in molecular biology: flow cytometry protocols, 2nd edn. Humana Press, Totowa, NJ, pp 33–43
16. Montemayor M, Galofré B, Ribas F, Lucena F (2007) Comparative study between two laser scanning cytometers and epifluorescence microscopy for detection of *Cryptosporidium* oocysts in water. Cytometry A 71:173–179
17. Pina-Vaz C, Rodrigues AG (2010) Evaluation of antifungal susceptibility using flow cytometry. Methods Mol Biol 638:281–289
18. Pina-Vaz C, Rodrigues AG, Sansonetty F, Martinez-De-Oliveira J, Fonseca AF, Mårdh P-A (2000) Antifungal activity of local anesthetics against *Candida* species. Infect Dis Obstet Gynecol 8:124–137
19. Pina-Vaz C, Sansonetty F, Rodrigues AG, Martinez-De-Oliveira J, Fonseca AF, Mårdh P-A (2000) Antifungal activity of ibuprofen alone and in combination with fluconazole against *Candida* species. J Med Microbiol 49:831–840
20. Pina-Vaz C, Sansonetty F, Rodrigues AG, Costa-Oliveira S, Tavares C, Martinezde-Oliveira J (2001) Cytometric approach for a rapid evaluation of susceptibility of *Candida* strains to antifungals. Clin Microbiol Infect 7:609–618
21. Pina-Vaz C, Costa-de-Oliveira S, Rodrigues AG, Espinel-Ingroff A (2005) Comparison of two probes for testing susceptibilities of pathogenic yeasts to voriconazole, itraconazole, and caspofungin by flow cytometry. J Clin Microbiol 43:4674–4679
22. Pina-Vaz C, Rodrigues AG, Costa-de-Oliveira S, Ricardo E, Mårdh P-A (2005) Potent synergic effect between ibuprofen and azoles on *Candida* resulting from blockade of efflux pumps as determined by FUN-1 staining and flow cytometry. J Antimicrob Chemother 56:678–685
23. Barbosa J, Bragada C, Costa-de-Oliveira S, Ricardo E, Rodrigues AG, Pina-Vaz C (2010) A new method for the detection of *Pneumocystis jirovecii* using flow cytometry. Eur J Clin Microbiol Infect Dis 29:1147–1152
24. Pina-Vaz C, Costa-Oliveira S, Rodrigues AG, Salvador A (2004) Novel method using a laser scanning cytometer for detection of mycobacteria in clinical samples. J Clin Microbiol 42:906–908
25. Pina-Vaz C, Costa-de-Oliveira S, Rodrigues AG (2005) Safe susceptibility testing of *Mycobacterium tuberculosis* by flow cytometry with the fluorescent nucleic acid stain SYTO 16. J Med Microbiol 54:77–81
26. Barbosa JM, Costa-de-Oliveira S, Rodrigues AG, Hanscheid T, Shapiro H, Pina- Vaz C (2008) A flow cytometric protocol for detection of *Cryptosporidium* spp. Cytometry A 73:44–47
27. Barbosa J, Costa-de-Oliveira S, Rodrigues AG, Pina-Vaz C (2008) Optimization of a flow cytometry protocol for detection and viability assessment of *Giardia lamblia*. Travel Med Infect Dis 6:234–239
28. Barbosa J, Rodrigues AG, Pina-Vaz C (2009) Cytometric approach for detection of *Encephalitozoon intestinalis*, an emergent agent. Clin Vaccine Immunol 16:1021–1024

Index

Louise O'Connor and Barry Glynn (eds.), *Fungal Diagnostics: Methods and Protocols*, Methods in Molecular Biology, vol. 968, DOI 10.1007/978-1-62703-257-5, © Springer Science+Business Media New York 2013

MIX
Papier aus verantwortungsvollen Quellen
Paper from responsible sources
FSC® C105338

If you have any concerns about our products,
you can contact us on
ProductSafety@springernature.com

In case Publisher is established outside the EU,
the EU authorized representative is:
Springer Nature Customer Service Center GmbH
Europaplatz 3, 69115 Heidelberg, Germany

Printed by Libri Plureos GmbH
in Hamburg, Germany